Bradovich

ALSO BY WILLIAM HERRICK

The Itinerant (1967)
Strayhorn, a Corrupt Among Mortals (1968)
Hermanos! (1969)
The Last to Die (1971)
Golcz (1976)
Shadows and Wolves (1980)
Love and Terror (1981)
Kill Memory (1983)
That's Life (1985)

Bradovich

a novel by
WILLIAM HERRICK

A New Directions Book

With many thanks to Yaddo.

Manufactured in the United States of America
New Directions books are printed on acid-free paper.
First published clothbound by New Directions in 1990
Published simultaneously in Canada by
Penguin Books Canada Limited

Library of Congress Cataloging-in-Publication Data

Herrick, William, 1915–
  Bradovich : a novel / by William Herrick.
     p.   cm.
  "A New Directions book."
  ISBN 0–8112–1141–X (alk. paper) : $18.95
  I. Title.
PS3558.E75B7   1990
813'.54—dc20   90–5874
                                     CIP

New Directions Books are published for James Laughlin
by New Directions Publishing Corporation,
80 Eighth Avenue, New York 10011

For Sarah Herrick Morris

Bradovich

*B*radovich. He was a Croat from Chicago. "Don't call me a Hunkie and don't call me a Polack. My father's parents came from Obradovich, Croatia. And that was their name—Obradovich. An illiterate immigration official on Ellis Island said, 'Jesus, you're not Irish,' and dropped the O."

Stephen Bradovich—Brad when he played tackle in high school and then for U. of Michigan. Wizard Bradovich when he played defensive end for the Raiders—Wizard because there was no offensive trap he couldn't spring to sack the quarterback. He played with the Raiders for six years, was all-pro a couple of times, then busted his knee. Gave it up. "I wasn't going to become a cripple for life just to give fat cats a thrill on boring Sunday afternoons."

Bradovich was a gnarled oak of a man, a muscle, a hard knuckle-scarred fist. He wore a size 52, long. Was in top shape, ran or walked a brisk five, six miles twice a week, winter or summer.

When he was asked why, where was he going at his age, he answered, "Have to be in shape when I meet my Maker—he'll sure as hell send me to the hot place." Then he laughed, the scarred face and bashed nose rearranging themselves. City snobs stared at his lined face, his partially flattened nose, his sheer bulk and assumed he was barely literate. Bradovich was graduated from Michigan with honors. He knew how to read and the difference between history and the smell of skunk. He could tell you who he'd cribbed that from, too. "Rebecca West. A very smart lady. Knew my old folks' country better than they did."

*B*radovich was a sculptor now, chiseled faces from stone. Faces blanker and blanker. Soon there would be no distinguishing features. No eyes, no nose, no mouth, just an oval with incised ears. Perhaps a hairlock like a Mohawk's. A blank face for all the blank faces out there.

Some critic—and to Bradovich all critics were asses—asked, "What statement are you making? What truth are you seeking?"

Bradovich laughed. "If you hear a man squeaking about the truth, you know immediately it's his own truth he wants to hear about, not yours or anyone else's. The booers and hissers of the right/left, the violent pushers of the bullshit, the frantic dancers of the ideological rag, ta ra, ta ra! You kill me and I'll kill you, ha ha, ha ha! Me, I'm an anarchist—a pacifist. Just raise your fist at me and I'll slaughter you." He laughed again. "See. I'm a hypocrite like all the rest. What truth? Whose truth? Yours? The mob's? Screw you, the whole lot of you. I chisel my faces blanker and blanker. Where do I come from? My uncircumcised father,

4

my unJewed mother, my blond reed of an uncle, the epitome of Western man. (You see, I can be as pretentious as the next man.) I am their son. I am a free man."

It was during his stay in the hospital having his busted knee repaired that he discovered a use for his hands other than pounding opposing linemen. A social worker brought him molding clay to fool with as he lay in bed his knee immobilized, and with an excitement bordering on exhilaration Bradovich found himself fathering a population of heads—grinning faces of nurses, doctors, other patients. He was accepted as a student at the Art Institute of Chicago. "You have talent," they told him. "Sure," he said. "Why not?"

With big knuckle-scarred hands he required a big medium. Stone. He loved stone. Marble. Granite. "A hammer and chisel. Just like my old man."

Playing pro ball, he'd saved his money, had enough to live on comfortably. Went mostly with decent women—avoided the avid whores who hung around professional ballplayers. He had a normal appetite. "When I do it, I enjoy it. Fucking's a riot, isn't it? A sight gag. There she is, flat on her back, agape, her eyes glazed, and there you are, on your knees, chest heaving, mouth salivating, your jut jutting, and you're so much in a hurry you miss what's wide open, stab here, stab there, until she grabs it with impatient fist and leads the half-blind fool home. All done with high seriousness, when it has to be the most ludicrous sight on earth. Every time I think about it, I laugh like hell. But if you get to laughing when you're at it, you lose the damn thing, and she's sore and you're so unnerved you want to smash the wall."

At art school he met a woman he thought he loved. They married and came to New York. "If you're going to make it," she said, "that's the place to do it."

5

She was a painfully thin woman, with long skinny legs, a small behind and huge breasts. If her face had been just a bit filled out, she might have been said to be pretty, but as it was her chin was too pointed, her nose a mite too sharp. It was her eyes which drew him—they were large, vibrant, dark, sparkling with intelligence and wit. Sharp, quick. If only she'd been blessed with wisdom, too. Her movements were so staccato you got the impression at any moment she might slip into hysteria and her brittle high-pitched voice would snap into fragments, splinter like fragile old glass.

Frances, her name was, Frankie. But he was no Johnny. She dominated every dinner table posh or poor, her voice cutting through all others so that out of an inhibition against shouting they had to shut up and yield to her pervasive chatter and laugh.

"I'm a vaudevillian," she told him. "Must do whatever's necessary to capture center stage, otherwise I become scared. Feel all alone, shunted off into a corner, neglected." She was also mean-spirited to the point Bradovich would want to strangle her. Little things. Managing to put down her friends to the benefit of herself. Telling them she and he were invited to parties she knew they weren't. Had a continuous and endless need to pat herself cutely on the back. Bradovich would close his eyes and see her as a little ten year old in patent leather party shoes, white socks to beneath her knees, a big bow on her head, a bigger bow tied around her waist, knee inclined: "Aren't I the prettiest girl here? Aren't I the smartest?" She would never show you her painting and wait for your honest comment, but would show it to you and say, "Isn't it a gorgeous work of art, now isn't it?" What could you answer unless you wanted to destroy her?

Bradovich uttered the same words when they separated as he uttered when he first met her. "Poor kid."

He lived with her a short time, then married her, then

learned to hate her. That sharp voice, that cutting tongue, that narrow meanness. That godawful fear. "I couldn't get up the guts to leave her," he told Valerian his second wife, "and was as happy as a kid with an ice cream cone when she fell for a famous painter she met in my studio and left me for him. Prick that he was and is, they were right for each other. All these years later and they're still together, except she's become the famous painter and he the bullshit artist."

Bradovich—most didn't even know his first name. Everyone called him Bradovich, some Brad, a few Wizard, his old football name. When his kids got fresh they also called him Bradovich. Valerian called him Mister Bradovich, a family joke. He signed his work Bradovich, and when he had an exhibit it was billed Bradovich. He loved his name, it fit him like his skin, tough and scarred.

His second wife answered his ad for a model, and stayed with him for almost thirty years—model, lover, wife. Then she died of cancer. Went very fast. They'd been a close couple; it was obvious to all who knew them that they'd never stopped loving each other. "As hot for her the last time as the first," he said aloud to no one in particular after she was gone. That was a habit of his, to talk aloud to no one in particular. He took Valerian's death very hard.

"She was a wise, good woman," he told Floriana, his friend. "Those are the ones who rack it up first, aren't they? I've done every bloody evil thing a man can do, and here I am. Still could, if put to it, tear up a quarterback's spine."

Valerian and he had two children, a boy and a girl. Martin and Laura. In their early thirties now, they were both married, and each had a child. Bradovich was a grandfather. He loved his grandchildren, though he'd never seen them.

7

His children and their families lived on the Pacific, and he hadn't been able to get away to visit for a long time. He was always busy preparing for a show. "They're toddlers, I'll get away, I'll see them soon." He sounded apologetic—was Bradovich mellowing? "Life just works out that way. My children keep asking me to move out there, to live with them, whatever. They phone me at least every week, they know I still miss their mother, Valerian—her father was one of those roots, fruits and noots freaks. Never ate beef, fish or fowl. A health nut. Died of a heart attack at age forty-eight. Val. A wise woman. 'Slow down, soften up, Mister Bradovich, or you'll have a hard end. Too hard even for a man who chisels stone.'

"Yup, she was right, it's getting harder. Look, my children—I've always given them space, they're giving me mine. One of these days I'll close down the studio, triple lock my apartment, and fly out there. Will probably never return. Live out my old age in the sun with my grandchildren climbing all over me." He looked away.

*B*radovich's studio was downtown in the old warehouse district. It used to be a stable and wagonhouse for a beer distributor before Prohibition. Then it became a truck garage. It had a flat roof into which he had installed skylights. A caretaker who guarded several of the neighboring buildings looked after it for him. When Wizard found it, it was vacant, had been for many years. "I can still smell the horses—is there anything sweeter than horse shit? Now the area's gentrified, taken over by artists who've *made it* and their glitzy art galleries. I don't pay them any attention. I do my work. Have enough commissions to keep me busy till the day I croak. Some people ask, how much work does it take to make a blank face? I laugh at them. Jerks, philistines, looking for kitsch. Go to the subway, I tell them, you'll see great junk. The decline and fall. I suppose I'm straight. I work for a living—it just turns out I love my work more than most. And it's not therapy either. You need therapy, go see a psychiatrist." Bradovich laughed, more like a sneer. "I never boozed, never poked around looking for pussy, never pretended I was living in the Latin Quarter. After work I went home to my wife and kids. I'm an artist, not a bohemian. Never wore a beret in my life. Ha."

Blank oval faces. Polished stone. After his wife died— "I'm glad I'm going fast, darling, the pain's just awful. If it takes too long, you'll do it for me, won't you?" "I promise. I won't let you suffer, Val, you can bet on it. I love you. It's been marvelous." After she died, he stayed longer hours at the studio, frequently worked around the clock. Polished marble heads. Not a blemish. Beautiful blank polished faces. That was his trademark, his style, his voice, his name. Sometimes the oval was elongated, at other times flattened. On occasion when he found the right piece of wood he'd do a wood head. But these he painted. Eyes caught wide open in surprise or fear, white face, black face, mouth agape, you could hear the words, "Oh, my God!"

Every day was a new surprise. One day you were laughing, the next bawling your head off. If you let it get to you, you ran wild and then everything was the same, identical days seen through a haze, a boozy fog.

Bradovich lived uptown on the west side, in the same apartment in which his children had grown up, and in which Valerian had died. It was on the ninth floor: the front looked out over the boulevard, the rear into a narrow rarely sunlit court. It had a triple lock on the front door. Downstairs in the lobby stood a doorman who never opened the door for anyone, merely guarded it. Someone to say good morning and good night to. He was called Slim, though his days of slimness were long gone. He had a sour face, broad shoulders, and a paunch. Slim was trustworthy. He had keys to everyone's apartment and no one had ever lost a thing.

Whatever it was began in Bradovich's studio. It was a winter day, cold, the wind a knife. Facing him from around the perimeter of his studio were many of his heads, some unfinished, others not quite. Stone mute, eyeless. Wood round-eyed, agape. He was polishing a head commissioned by a museum in Rome. The head was clamped in a vise, its grips padded, of his own invention. He was using a hand held rotary polisher with the finest abrasive. Someone was looking over his shoulder. He could actually feel it. Nerves. Just nerves. As he delicately rotated the polisher over the piece, green granite, flat head, long chin, he kept turning to look over his shoulder to see whoever or whatever it was and would scar his work, almost imperceptibly. The results were not what he wanted, and he kept having to obliterate the scars, changing the diamond abrasive frequently.

Absolutely no one and nothing was peering over Bradovich's shoulder. Tired. Burned out. Time to close shop.

11

Get out of here. Stay home. Pay Floriana a visit. Have a beer with Golo. Enough of this. Tired. Very tired. Still, whoever it was persisted in staring over Bradovich's shoulder, observed every move he made. With great effort, he willed himself to concentrate on his work. "Fuck it," he said aloud. Those two words said it all for Bradovich. Always had.

Finally, he finished the piece, crated it, steel-banded it, addressed it with black india ink, signed the shipping papers, and had the freight forwarder pick it up. "Goodbye and good luck," he said when he saw the last of it.

He cleaned and oiled the polishing machine, fitted its cloth hood over it. Every tool in place, the floor swept, he set the alarm, fought his way into his heavy mackinaw, took one last look over the studio, punched the lights out, bolted the heavy doors behind him with two steel bars. Whoever or whatever had been staring over his shoulders had left him. He smiled to himself. Stir crazy.

He was alone now in the old cobblestoned street except for the usual pedestrians, coat collars up around their ears, hats pulled down, shoulders hunched against the cold, teeth clenched behind frozen lips.

The subway platform was an icebox, and the body heat of the mob inside the subway car pleasantly acceptable for a change.

On the boulevard, three blocks south of the apartment house, Bradovich stopped in a grocery to buy several items for his dinner. It was cold in the store.

"Why don't you get this place heated, for God's sake?"

"None of your bloody business," said the owner-clerk. They had known each other for some fifteen years.

"Miser. Some day we'll have to chop you out of the ice."

"God bless, Brad."

"*Con dios*," he said as he left.

The broad avenue, divided through its long winding length by an island of litter and half-dead bushes, was frostily lit, and the faces of the pedestrians reflected a gelid whiteness. He merely lowered his head against the wind and forged ahead to his destination. Out of the corner of his eye he caught sight of a multi-colored shawl flapping about, and he raised his eyes to see a young woman draped to the nose, and though her face was half hidden he thought he recognized an actress then starring in an exotic theatrical production. He almost wolf-whistled, then remembered his age and the time. Floriana, here I come, I am in need.

Slimless Slim, the doorman, more sour-faced than usual, greeted him with a grunt to which Bradovich responded with a raised shoulder. It was the weather, nasty and mean. There was no mail in the box except junk which he flipped into the bin kept there for that purpose. The elevator rose slowly, clanking and rattling, a few times hesitating, at any moment threatening to stop at midfloor. Wizard merely leaned against the iron grille wall, having full confidence that the old lift would not betray him. At the ninth floor it groaned to a stop, the door clankily slid open, and he stepped out. Shower, then eat. Hell, no, eat first. Hungry, very hungry.

He slid the key into the special triple lock guaranteed foolproof against tampering. It was, to his astonishment, open. The door swung free as if by the force of its own weight. "Damn!" He dropped his package, came in fists first, shoulders hunched in a fighter's stance. The dim hall light was on and a tall, square-shouldered man with what looked like a face chipped from stone stood staring at him. As Bradovich came at the man, a second, a twin to the first, introduced the tip of a long butcher knife to Bradovich's gullet, then forcefully pushed him against the wall.

"Don't even think of it, Mr. B.," the first man said calmly. They apparently knew him.

Bradovich was no fool. "My wallet's in my back left pocket. Take it and leave." His voice was controlled. Never show panic to these guys, he always said. He didn't now.

Two hard impassive faces stared at him.

The first said, "We're not here to rob you, that's against the rules. You are under surveillance, pending review, that's the message The Authority sends you."

"What authority? Whose review?"

"You have been given the message."

"Who? What?" Bradovich shouted, the point of the knife pinking the skin of his gullet.

"That's not for us to say. We merely do our duty."

Bradovich was breathing heavily, at every breath the skin of his adam's apple punctured a bit more, and the knife blade was beginning to show blood. His impulse, natural for him, was to go at these men, to take what came. But he knew the knife tip would puncture his gullet before he could bring up knee or fist.

He remembered his rights. "Where's your warrant?"

They both smiled thinly.

"I demand the right to call my lawyer."

The twins shrugged their broad square shoulders in unison. "You'll be wasting your time," the first said, then nodded to his brother. Slowly, led by the unarmed twin, they backed out of the door, the knife held at the ready. The knife wielder slammed the door behind him and it automatically locked. By the time Bradovich's angry fingers managed to open it, and he was able to run into the outside corridor, the men were gone. How, he didn't know, since he heard no footsteps on the stairs and the elevator was not moving.

Bradovich reentered his apartment and raised the house phone which hung near the front door. He rang for the doorman. Slim didn't answer immediately, and when he did he said the two tall men had already left. "Just saw their backs, never caught a gander at their faces."

"How'd they get in in the first place?"

"Tell you the truth, Brad, I don't remember ever letting them in."

"Good for you!"

Slim banged the phone down.

So did Bradovich.

After washing the thinly caked blood from his gullet, he called the police from the kitchen phone. Two officers, a man and a woman, appeared at his door in about ten minutes. He explained what had happened. They examined the door lock, an intricate steel mechanism.

The woman officer, stocky, deep-voiced, peroxide blonde (perhaps a transexual?) said, "No one's tampered with the lock. Whoever opened it had a key. Who else has one besides you?"

"Just Slim downstairs, but he didn't let them in, he never even saw them come in."

"Are you sure about him? He looks—"

"Looks are deceiving. He's rock honest—and as tough as they come."

"Okay, okay. Was it your kitchen knife he used?"

"No. Mine's still in its slot."

He then gave the officers a description of the men who had manhandled him, and they prepared to leave, saying they would check out the files and inform the computer.

"Thanks."

Before leaving, the female cop stepped close—smelled of Chanel No. 5, Val's favorite—without asking lifted his chin, and examined the wound with thick fingers. "Superficial. Put some iodine on it to make sure it doesn't become infected."

"Sure, thanks."

Bradovich locked the door after them, chaining it as well. He stood leaning with his back against it. This had to be a gag, some wise bastards thinking they were going to have some fun at his expense. But who? When he played ball there'd been feuds with opposing players, but that had been a very long time ago; he couldn't even remember their names anymore. There were some artists he'd bad-mouthed. Kitsch. Phony. Glib. Junk. Hot air. Academic. Those were the words he loved to use to describe so many of them. Most artists were thin-skinned, including himself. Yeah, that must be it. One of those new wiseass kids. Schlagel. Palle. Whatsisname, the great genius, épating the bourgeoisie—a great piece of work, a kid looking up his mother's cunt. Wow! What a statement! Fuck 'em. They want to play games, I'll wrack them up.

The stonefaced twins would be back, he knew it in his bones. Under surveillance—bull. When he had told the cops what the men had said, they'd looked at one another knowingly, as if they'd heard this story before. When he questioned them about whether they had, they denied it. He had the distinct feeling they were a little frightened themselves. It couldn't be. An hallucination perhaps. No, it wasn't an hallucination, don't start that crap. The lock had been open. Slim had seen them as they left. That wasn't ketchup on his gullet. The knife blade hadn't been made of rubber. The hell with it; he was hungry.

He went to the toilet and applied iodine to his wound. It wasn't much. Then he relieved the pressure on his bladder. So nervous was he from his strange encounter that he splashed the tile floor before settling the stream into its required course. A fastidious man, he washed the tiles clean

16

with an old sponge which he kept handy for such purposes. He smiled as he did it. Val would be proud. She'd always been after him to clean up his piss drops and not leave them for her. He then soaped and washed his face and hands. His hands were calm now, without a tremor. "Go get 'em, Wizard!" He laughed.

The kitchen was efficient, but small, so small that when Val was alive only one of them could work in it at a time. Only when the mood was upon them would they work in it together. Every time they turned, they copped a feel. Once they made love on the floor, and it was like making love in a crate. A mouse had peeked out from under the refrigerator, seen them and scampered back. "Your pussy scared him away, honey."

She slapped his face. "Don't you ever dare call it a pussy. Call it what it is. A cunt."

They laughed so hard they had to give up all thought of fucking.

Bradovich put coffee on to brew, sliced a tomato, buttered two hard rolls, cut an onion into quarters, and sat down in the dinette to eat the smoked trout he'd bought for himself. Since that delicacy was among his favorite dishes, not even the perplexing occurrence could mar his pleasure. He was thick-headed—could do only one thing at a time. If he was scared, he couldn't eat. If he ate, he couldn't be scared. He downed fish, rolls, tomato, and onion to their very end, leaving his plate so empty no one could possibly have discerned he'd eaten from it.

He refilled his cup and sat slowly sipping at it. What had he done and to whom? It was a dirty trick. If he were a weaker man, it could have given him a heart attack. A man lived a lifetime in a day. A man was not an angel. Neither was he a saint. He'd had his innings. Broken a couple of noses. Stepped on a lot of toes. It was true, he was known as a

17

tough, contentious bastard. He hadn't been a vegetable. "You're alive, my son, know how to fight."

"Yes, Mom."

She'd come alive, too, after his father died of a heart attack. In late middle age, she'd fucked every Polack, Hunky and Croat on the block. "What's got into you, Mom?"

"I don't know. It's too bad I'm clean, though."

"What do you mean by that?"

"I'd be happy to give all of them V.D. Jewhaters, the lot of them."

"Not all, Mom, just most."

She'd laughed. From mouse to tigress. Strange.

He'd gotten her a decent apartment overlooking the lake. In old age she began to thrive. Became chic—stylish, as she called it. She and Uncle Walt. The opera, the ballet, the symphony, read Saul Bellow, Chicago's very own. Yes, no day's like yesterday. Who knows about tomorrow? She was gone now. Uncle Walt, too. Val was gone, died in her youth as far as he was concerned.

Bradovich had one sibling, a sister, Sybil. A tall, retroussé-nosed green-eyed blond. Kept on the move. Independent, an alley cat, tough. Laughed a lot. Threw tarot cards, danced in a chorus line. Now made stained-glass windows for ugly concrete churches. He hadn't heard from her in years; they had never gotten along well.

"How come only two, Mom?"

"Your father wanted a brood, but I didn't. I used a rubber derby until we went to New York on a trip. I slipped away one afternoon and went to the Sanger Clinic. They put in a device—I suppose an early I.U.D. After a while, your father stopped talking about more kids. The I.U.D. never hurt me, so I came out okay. You and Sybil were wild. Your father was wild, too, and never grew up, so it was like I had a dozen. Too much. I was a mouse, wasn't I?"

"What makes the difference now?"

"When he died, I got rid of a stone, a weight, that—"

"I don't want to hear about it. He was my father. I loved him just as I love you."

She dropped it. Never brought it up again. Mothers and fathers, they never leave you. It's only when you come to terms with them that you grow up. Sometimes it doesn't happen until you're fifty. He ought to go call his kids. Was too tired. Those stoney-faced bastards. . . .

Bradovich stood up from the table, washed his dishes, swept up the few crumbs, strode to the living room to relax, to finish reading the daily paper which had been left at his door before breakfast that morning. The living room was Val's—sort of near eastern, with a large oriental rug on the floor, hooked hangings in maroons, reds and burnt oranges on the walls, a good solid large sofa in champagne-colored upholstery and a pile of pillows to match, several paintings from friends with whom Bradovich had swapped, and a Bradovich blond marble head of Valerian—realistic, the surface rough—on a black marble pedestal in the corner. The easy chairs, both large and luxuriously soft, were old and comfortably worn. He sat in the one near the window under the original Tiffany lamp, and settled the paper in his lap. It was open to an inner page on which was recounted the news story of a wealthy young man who was accused of murdering and dismembering his wife because she had slept with a man whom he, her husband, had seduced into her bed. After a passionate plea by Bennett Pollack, considered by the cognoscenti to be the finest criminal lawyer in the city, the accused had been released on his own recognizance, pending further investigation by the police.

Bradovich trembled. The words "released on his own recognizance, pending further investigation" were underlined in heavy lead pencil. He had not done this that morning.

Then who had? Those two sons of bitches. Surveillance, they had said. Pending review. What review? Bradovich was now standing, shaking with fury. Balling the newspaper in his hand, he threw it underfoot and stamped on it.

The phone rang. He hurried to the kitchen to answer it. It was the low bass of the female police officer who had come to the apartment to investigate his complaint. "The computer indicates negative on you, sir. No security force in the country has you listed. Neither does Interpol. If you are bothered again, let us know."

"Negative. Thanks."

"You're welcome, sir. Have a good night."

If they weren't official, then who the hell were they? He remembered the business with someone looking over his shoulder all day in the studio. Crazy. The whole thing was nuts. Slim had seen them. And he'd never underlined a word in a newspaper in his life.

Bradovich began to know fear. He'd known it before and knew exactly what to do with it.

He put on a blue sweatshirt, then a heavy wool sweater over that, and went down to the street. The wind had stopped gusting, the air was cold crisp, the night sky blue black, the moon quartered, the stars heavily concentrated overhead. The unsheltered sprawled in their own frozen piss in the doorways.

He marched at a fast clip south for twenty blocks. Whoever they were, they were not going to have an easy time of it. He turned north, still at a fast pace. By a turn of his eyes he could see those who walked behind him reflected in the large store windows. Be alert! Someone was following him. At the next corner, against the light, he abruptly turned and ran across the vehicle-filled street. Resumed his march northward. Shortly, he saw a tall square-shouldered figure reflected in the store window to his right. He continued at his fast gait, then stopped in his tracks, spun quickly, his hands fisted. The man was nowhere present, just an elderly woman who had stopped in time to avoid bumping into him.

False alarm. Maybe. Or were they up to using tricks to upset him? Well, he knew quite a few tricks himself. "I can take care of myself," he said to no one in particular.

As he neared his apartment building, he met a woman of his acquaintance who also lived in the building. She had an armful of packages—he'd rarely seen her when she hadn't—and he relieved her of two and walked alongside her chatting about the weather. Her name was Beatrice Holden. She always made the same joke. "I am beholden." The first two times he'd smiled. After that he merely ignored it.

She was a tall, slender woman, though full at hip and bosom. Her curly red hair flecked with gray floated above her head like a puff of cloud. You expected that before she entered a doorway the cloud would float off into the sky and that indoors she would walk around baldheaded. Beatrice's face was freckled and her nose pointed like a carrot stick. She was pretty in a vague sort of way. Prettier thought about than actually looked at. Bradovich was fond of her because she was generous and seemed to favor him. After he and his children returned from Val's funeral, they found the apartment spotless and all sorts of deli dishes and wines and sodas laid out on the dining-room table for the grieving family and friends. Wizard's painter and sculptor colleagues had wanted a real bash. "You oughtta get drunk, Bradovich," but Val had always hated those drunken binges, so why should he demean her memory the day of her funeral? Beatrice had done it as Val would have. That time it was he who said to Beatrice, "I am beholden." She smiled gently and kissed him

21

on his cheek. She was divorced, had two children who were graduate students somewhere. Was a good five years younger than Val, which made her fifteen years younger than he. Bradovich kept waiting for the right hint to jump into her bed. But it was never forthcoming.

Usually Beatrice Holden was quite talkative whenever they met. Tonight, however, she was laconic, stopped before the entrance to their building, thanked him, reached over to take her packages. He offered to help her to her apartment, or the elevator at least, but she declined with a little smile.

What the hell was eating her? He wondered if she knew about what had happened. Slim gossiped a lot. Forget it! You're beginning to act like an old crock.

He decided to continue his march northward for another twenty blocks. Though he kept up a fast forced pace, he was alert to those about him. Jealousy struck sharply as he observed a beautiful young black couple, both garbed in white jogging suits, their skin gleaming under the street lights, walking close together, their arms twined about each other's waists. At the corner they stopped to kiss. He remembered suddenly the ferocious hungers of his adolescence. It was good that the girls he went with knew how to ward off tough kids like himself. In those days girls didn't hike up their skirts that fast. Besides, lots of them had papas who made them spread their legs on the kitchen table under a bright light for an examination of their hymens. If anything was child abuse, that was. All in the name of godly chastity. . . . Sally. . . . Why did he remember Sally now? She was as wild and as tough as he was. And after they started getting laid every chance they had and her father demanded she get under those kitchen lights, she hauled off and whacked him one. "I'll get Brad after you. He'll beat the living shit out of you, Pa." She knew the old man just wanted a smell of that sweet pussy of hers—the sweetest on the block.

An old woman clutched at his sleeve. She was swathed in layers of dresses, sweaters, scarves, all wrapped about herself helter skelter, stained by every soup she'd ever slurped, every bit of vomit retched. She asked for a dollar, and he knew it was for booze. On other days he would have brushed past her, but today he hesitated, then withdrew a dollar from his trouser pocket. She seized it with bony claws and without a word rushed on to the next pedestrian. Again successful, she sped on to yet a third. As Bradovich pushed by her, she grabbed his sleeve again, but this time he ignored her. She either had an insatiable thirst or supported a duplex apartment in the swanky part of town.

By the end of this twenty block forced march, Bradovich had worked off his fear. The whole business was probably some sort of mistake, it would straighten out somehow, and he slowed his pace as he ambled back towards his apartment house. Why hang around the city? Winter had come to stay. The studio was locked. His work was done. Not an idea in his head. Go visit your kids, see your grandchildren. What's holding you now? One thought led to another and he was suddenly attracted by the resort clothes in a nearby store window. He stopped to window shop and as he did saw reflected in the glass a tall square-shouldered figure. He rotated on a heel, looping a powerful uppercut at what he hoped would be the man's jaw, but all he hit was empty space. Bradovich caught sight of the man walking quickly south and set off after him. If I catch him he'll be ready for the hospital. Just as he neared the man who'd been shadowing him, someone from out of nowhere threw a professional block at him. Hit below the hips, he went down on his ass. Before he could climb to his feet, the two had disappeared.

Bradovich stood in the middle of the sidewalk, a big hulk of a man, his huge fists on his hips, staring into space. Now he knew what murderous rage actually meant. "If I catch those guys, I'll kill them," he muttered aloud. "Kill, not maim, not bloody, kill!"

Infuriated as he was—scared as well though he wouldn't admit it to himself—he hurried toward the safety of his building. As he approached its entrance, he saw Beatrice Holden down the block speaking to a tall man as they both looked in his direction. When they saw him staring at them, they turned quickly away. On impulse, he stepped toward them belligerently, but changed his mind and quickly entered the lobby.

Bradovich needed comforting, so he had the elevator stop on the sixth floor, where he strode quickly to apartment 6C, rang the bell, one short and one long. In a few seconds he was aware of being examined by an eye in the aperture. Then he heard a clanking of bolts, followed by the opening of the door the length of a chain secured to both door and doorframe. Floriana's plump pretty face stared at him with deliquescent eyes.

"I can see you're upset, Bradovich, but I'm sorry, it's impossible this evening."

As she hadn't closed the door in his face, but stood peering at him, he had the opportunity to say, "Tell me, Flor, could there be a reason other than the fact you're busy for turning me down?"

Floriana smiled to reveal very white but irregular teeth. A tiny vein throbbed in her upper lip. The red tip of her tongue wet her lips. "Though I've heard a vague rumor about your problem"—his heart fell—"I assure you, Wizard, I wouldn't let that interfere. What about tomorrow at 9?"

"Okay, thanks, Flor."

Nodding to him, a secret smile scribing her voluptuous lips, Floriana closed the door, and he could hear her securing the many devices which protected her from mayhem and theft.

He approached his door anticipating trouble, but found everything in its proper order. After taking a shower, he was glad to get to bed. The light out, snugly wrapped in his blanket, he remembered he had not secured his apartment door properly. Senility was setting in. He resisted leaving the comfort of the bed. No, no, better get up, you slug. It occurred to him as he bolted and chained the door that he might have forgotten to do so before he'd left for the studio that morning. No way. They're getting to you, Bradovich. If only he could grab hold of one of them. Just once. For ten seconds.

Bradovich lay under the blanket, his eyes shut, concentrating totally on defeating fear. He slowed the pace of his breathing, relaxed his body, his arms, legs. He fell asleep.

As Bradovich was brushing his short gray black hair in the morning, the phone rang. Jesus, it's not even 7:30.

"Yes?"

"You will be prepared to make yourself available on short notice," a voice, male, said sharply. It had a metallic ring.

"Who the hell are you?"

"You are under surveillance, and it will go better for you if what occurred last night in the street is not repeated," the voice rasped. "Learn to hold your bad temper, Mr. B." The wire went dead before Bradovich could shout another word.

Bradovich felt as if he'd taken a powerful jab to the stomach and he was forced to sit down on the bed. As sweat beaded his forehead, he urged himself to slow his breathing. He was in A-one shape, why did he feel so damned weak? Who were these bastards? Why were they trying to do him

26

in? What the hell had he done? The police computer covered every security authority in the country, including the FBI and CIA, was wired to every major police force in the world, including the KGB, and it had come up negative on him.

So who were they? What had he done? Nothing, just lived. Got up in the morning. Washed. Dressed. Had breakfast. Straightened out his place. Into his coat. Subwayed to work. On good days, walked. Worked hard all day as most artists did. Tried not to backbite his peers. Had to admit he failed most times. Tried not to be envious of their successes, though he was jealous as hell. Since Val died, he paid Floriana a visit on occasion, they were pretty good friends. Went to a bash once in a while, flew to Paris, Rome, Barcelona to see what the competition was doing. Hopped over to Africa, too, to see what was left, hadn't been stolen. Primitive geniuses. Kept in close touch with his children. Yes, dammit, he had to visit them, to see his grandkids. Why, just why was he sitting there, tabulating the items of his life? Who the hell was he apologizing to? He'd done what he'd done and could never undo it. Said his *mea culpas* and fasted on *Yom Kippur*. What else can a man do? Pay, Bradovich, pay and pay. Christ, he wished Valerian the Wise were here. "Take it easy, Mister Bradovich, the worst that can happen to a man or woman is to die, and sometimes that's for the best. Don't exaggerate it, overblow it. Don't be the phony arteest?"

Bradovich laughed. That was the worst taunt. "Don't be a phony arteest, not like you've been blessed by a divine hand. If Leonardo da Vinci and Michelangelo didn't put on airs, why should anyone?" But this, my dear Val, isn't like that. There's some Authority after my ass.

He finished dressing, and went into his small kitchen where he had a breakfast of orange juice, whole wheat toast, a soft boiled egg, and black coffee. He noticed a tremor in his hands, but with just slight effort controlled them. All the symptoms of a pro before going into battle. Only a deadhead wouldn't have them.

The morning newspaper was outside his door. As he sat having a second cup of coffee, he read it. The same news as yesterday, except worse, and the names were different. A very sane world. A very rational species. Not even the home court advantage helped. "Why don't we give it up?" he said aloud as he stood, throwing the paper aside. Washed the dishes, swept the floor, though there wasn't a crumb. Decided to go to the park for a stroll. He was taking it easy, had lots of free time, ought to begin relaxing. What would be, would be. He grabbed a leather aviator's jacket from the closet, and left, not forgetting to lock the door behind him.

As he alighted from the elevator, Floriana entered from the street. She was dressed in a tight black jogging suit, wore a yellow band around her head to keep her long black hair in place. Since her profession required that she stay in excellent shape, Floriana went to the park every morning where she jogged a fast five miles. For safety, she carried a keen long blade in a leather holster strapped to her well-turned thigh. A man had attempted to attack her the past summer and Floriana had used her knife expertly and he'd required twenty stitches and a blood transfusion to keep him alive. Seeing Bradovich, Floriana paused, then peered at him through slitted eyes.

"What? What the hell you staring at?"

She shrugged, smiled gently. "You and Beatrice should get together, you're both lonely."

"Oh, stop the matchmaking, Flor. I've got other things on my mind."

"I understand, Wizard," and she did, besides she was a real friend. "We have a date tonight, be on time, Mr. B."

He followed her with his eyes as she sailed along on

her nifty legs to the elevator. "You should have been a dancer," he called.

She smiled at him over her shoulder, then blew him a kiss. He knew exactly what Floriana loved to hear.

The sky was cloudless, the air crisp and clear, not even the car fumes could mar it. The sun was a sharply defined disk. Aten. He walked toward the park several blocks east of the boulevard with long quick strides so that his body became pleasantly warm, and soon he could feel a moistness under his armpits. The sight of several young women dressed in various layers of what could only be clothes from an antique shop, giggling happily, slowed him down. Bradovich was an ogler. Valerian used to make sarcastic remarks to him about it. "You even take more time with the lingerie ads than with the news."

"I can't help it. I make no excuses, Val. I love women. Everything about them. Fat ones, skinny ones, in between. Do me something."

"When you're walking with me, at least do me the honor of paying strict attention to me. To me!"

"You're right, I'm sorry. I'll try."

"Till the next time."

Once in the park, Bradovich began a very brisk walk, almost but not quite a trot, around the cinder oval which circumscribed the lake. Men and women in expensive sweatsuits and jogging shoes were either running or walking in the ridiculous style of professional walkers. A few were acquaintances from the neighborhood, and they nodded as they passed. A quarter way around the oval a huge mastiff tore its leash loose from its master's wrist and came tearing at Bradovich, its mouth open, red tongue salivating and hanging from its maw with what looked like hope and anticipation of a good meal. For a moment Bradovich stood paralyzed, but, the great beast upon him, he sidestepped much like a toreador faced with a charging bull. As the dog went snarling past him, Wizard kicked it hard under its belly. He stared at the animal as it lay on the cinders yelping in pain.

Shortly, the dog's master, a tough-faced, muscular young man, came running up. "What the fuck d'ya do that for, Mister B.?"

He didn't know the tough, how come the tough knew him? "That kind of dog should be kept penned, or at least muzzled, buddy. And maybe you, too."

Bradovich was looking for a fight. If he beat up this young hoodlum, he'd feel better, especially since the man was wearing a black sweatshirt emblazoned in large red letters "T.A." The Authority, perhaps.

The tough squared off and Bradovich laughed—he was going to enjoy this. But there was a loud police whistle from the grass beyond the pipe rail fence, toward which both Bradovich and the young hood turned their heads. There stood one of the twins of the night before. Couldn't mistake him. Tall, square-shouldered, black suit, black shoes, white shirt, black tie, black fedora, and a gray topcoat. Stoneface. "Take care of your mutt, kid, I'm going to get that son-of-a-bitch."

Wizard ran swiftly toward The Authority's operative, but as he hurdled the low pipe rail fence, the man spun about and disappeared into the trees at the crest of a slight knoll. It was senseless to follow, and Bradovich turned his attention back to the young tough and his brute, but they, too, had by now disappeared. Maybe their aim was to drive him crazy, not to scare him. And he'd be damned if they weren't succeeding. Those about him were giving him the eye and shaking their heads. Embarrassed, he continued his stroll, now at a slow pace. His hands were trembling again, and he shoved them deep into his jacket pockets to control them.

The good follows upon the bad. Ask Pangloss. Bradovich smiled. There was his friend Golo hopping along on his skinny bowlegs. Golo was a little man, about four foot six (he had a twin who was a dwarf), one of the great clowns who performed in the major hippodromes of the world. Golo the Gimp, he was called, not that he was lame but because he had a strange little hop to his walk. He wooed the audiences of the world, and won them hands down.

No sooner did Golo and Bradovich shake hands than they moved to a nearby bench at the periphery of the park and sat down. Sitting, Golo's shortness became less noticeable, for his torso was almost normal in length, it was his lower limbs which were foreshortened, and Golo sought every opportunity to sit so that he wouldn't feel at a disadvantage when speaking with someone not of his own height, especially the mountainous Bradovich, for Golo feared nothing more than to appear ludicrous to anyone observing him. Golo, whose profession and craft it was to make people laugh, abhorred a sight gag when he wasn't performing. "They seem compelled to laugh just by staring at me—well, I have to say I'm not a figure of merriment and I find it distasteful to be laughed at merely for standing in the street."

To which Bradovich had once responded, "Unless you're being paid for it."

Golo had laughed. "You're right as usual, Wizard." Golo was no prig.

He was so generous a man and so sensitive to another's pains and miseries that his obsessive fear of looking ludicrous was ignored by his many friends, and he was muchloved. There were those who pretended to have affection for him because he was world famous, a man with great power, generous with his money, but Golo the Gimp was no fool, and neither was he sentimental, for he had an almost magical ability to read minds. "I can smell a phony at a hundred miles," he boasted, and lived up to his boast every time. There was nothing better Golo and Bradovich liked doing than going one up on who knew more phonies in the entertainment and art worlds.

As they sat down on the bench, Bradovich thought he'd amuse his friend with a joke about a fake artist he'd heard about, a man who after he finished his canvas turned his back on it and threw handfuls of silver and gold glitter at it over his shoulders, but before Wizard could say a word, Golo poked him in the chest, and said, "You're in some sort of trouble, aren't you?"

Bradovich just stared at him.

"It's no secret, I've heard about it from others. Beatrice Holden—by the way, she's nuts about you—who I met on her way to shop mentioned it, and Floriana when I visited her early this morning before she went out for her run. If you want to unload it on me, go ahead, I have strong shoulders." He smiled, his face wrinkling and his canines—another mistake of nature, most of his teeth seemed to be canines so when he smiled broadly it looked as if he were a grinning dog—gleaming whitely in the sun. "Say something, for God's sake." Bradovich's face had frozen, his ice blues staring blankly. "What's wrong, am I being nosey or what?"

Bradovich was staring at two tall men, the twins, at the corner, near the discount pharmacy there, talking heatedly to one another. Suddenly they looked at him, spat, and hurried away.

Yes, why not talk to Golo about it? He saw his friend peering at him with half-closed eyes, trying to read his mind.

Golo was not only a good friend, he had influential acquaintances throughout the world, people in powerful places, some said even in heaven. They weren't joking, they meant it. He had been able to interrupt a long harangue by the garrulous Fidel Castro to convince the dictator to release a Cuban acrobat who had been imprisoned for impersonating a gorilla smoking a large Havana cigar. Another time he had convinced General Pinochet to open the prison gates for a pianist who had once played a concert in the presidential palace for the inept Allende. Perhaps Golo could find a way to intervene on his behalf with The Authority.

So Bradovich turned his eyes to his friend and simply told him about the events of the previous day and of this morning, and the fact that he was here speaking to him, Golo, was only because he, Wizard Bradovich, had been released on his own recognizance, pending further investigation, and right at this very moment he was under surveillance, right here as he sat with Golo.

Greatly agitated, jiggling up and down on the bench, Golo exclaimed, "Damn you, do you mean to say you've already accepted it as fact?"

Bradovich suddenly realized what he had said. "No! No! I don't accept it. Someone's putting me on. I won't accept it. No!"

"That's better, old boy," Golo said, himself relaxing.

Abruptly, Golo leaped to his little feet, "Look, I have an important business engagement with my agent—a big offer, big bucks," and he laughed, canines shining in the morning sun. "By the way," he said shyly, "when you see Floriana tonight—she told me you had an appointment—tell her to at least consider my proposal. I want to marry that lovely woman, but she must give up her work, what would people think?" The latter merely rhetorical, he wasn't asking for Bradovich's advice.

"I'll make a real effort, but you don't have a chance, she loves her independence too much. You know what she'll say—it's my body, I'll do what I want with it. And as for Beatrice, you're dead wrong, she gave me the cold shoulder just last night."

Golo shrugged his narrow little shoulders and hopped off on his miniature bowed legs. He stopped abruptly, shouted back over his shoulder, "Never give in! Never!" Then he continued on his way.

*B*radovich headed back to the boulevard. It was Saturday morning and the neighborhood's young housewives—housewives only on Saturdays and Sundays, during the week they taxied or subwayed to places of employment downtown—were out doing their weekly shopping. He enjoyed ogling them in their tight jeans, fleshy clefts fore and aft. Since Valerian wasn't there to poke him in the arm, he could do it freely, without inhibition. On a rare occasion, a feisty young woman would exclaim, "What the hell you looking at, you dirty old man?"

He'd apologize, then say quietly, "Old but never dirty. I just enjoy looking at beauty, and to my eyes at least you are exceedingly so." What could she say to that?

He was glad he'd run into Golo. The Gimp was everyone's guru, swami, shrink, rabbi. Also counsel, interlocutor, father confessor. Perhaps even judge. Still, he'd known enough not to ask Bradovich what he might be guilty of.

And if Golo had asked, what would he have answered?

"Not guilty, no more than most. It's true, I'm not much in love with my fellow human beings, but I don't hold them in contempt—just don't like them. They do what comes easiest. In most ways, I do, too. I leave it to the smug self-righteous to sneer. If I call you a shit, at least I know I'm a shit, too. How many can admit to that?"

"What else, Wizard?" Golo would have asked.

"Screw off, Gimp, not now."

Golo would have pressed his thick lips over his dog teeth and shaken his big head with understanding.

Suddenly it had gotten very cold, and Bradovich raised the collar of his leather jacket. As he turned the corner onto the boulevard, two tall men with stony longjawed faces and hard blue eyes clasped him underneath his armpits. As he began to resist, he felt a knife point burning the skin of his back, and heard a raspy voice in his ear. "If you want to make a fuss about it, that knife will go right through you. Just relax, Mr. B.—you'll live longer that way." For emphasis the knife point stabbed a little deeper into Wizard's skin.

To make their job more difficult, Bradovich allowed his body to go limp, become a dead weight. He was a big man, let the bastards work. The knife point sank more deeply into his flesh, but he muttered, "If you want to kill me, go ahead, I'm not going to make it easy for you. And at least one of you cocksuckers will go with me." He compacted his muscular body and suddenly uncoiled violently, almost broke loose. The knife burned with a live flame as it tore flesh; the pain was excruciating, and Bradovich fainted.

When Wizard came to, he was lying on hard concrete, staring up at a roof of arched reinforced glass panels through which the sun shone with a diffused magenta light. Must be tinted glass, he mused, not quite conscious. He turned his head about and saw he was in a large empty space with whitewashed plaster walls. On one wall hung a huge poster

35

of a side of blood red beef hanging from a meat hook. Another was covered with an immense poster of a massive bulldog, its maw open to reveal a red cavern filled with long sharp canines, and he was reminded of Golo the Gimp.

As he squirmed about to view his prison better—was it a prison?—he realized the pain in his back was acute, but that he was tightly swathed in a bandage. The shirt he was wearing was not one of his, but of spotless white linen. His black leather aviator's jacket was lying near him, and he examined it. The hole in the back had been sewn neatly with tiny stitches that were barely noticeable. He could hear traffic from outside, and peering through a wide glass door, he could see pedestrians. Ha, the old A & P. It had been closed down for months. Maybe it's to be reopened again and I'm going to be cut up for the fresh meat counter. He heard the mewing of a cat and twisting his head saw a huge tom in a far corner playing with a mouse already paralyzed with fear. The cat batted the mouse back and forth, having great fun, a smug smile on its fat face.

Bradovich tried standing, but became dizzy, and stopped his exertions, just sat where he was. The place had a series of wide glass doors at one corner facing into the boulevard. If they were locked he would have to break the glass. This would probably set off an alarm, and he would be arrested by the cops for breaking and entering.

He closed his eyes for a moment—could hear the crunching of bones, the cat devouring the mouse—soon he'd recover his strength and be sent back into the game. It was the final game, the super bowl itself, you played even if it killed you. Rah rah.

Whoever The Authority was, its operatives were competent enough. The wound in his back did not hurt too much now, they must have cleaned it properly, patched it up, not maimed him too badly. What the hell? In a corner standing on a box was Madeleine Dearing, the great opera diva who'd married his friend Strayhorn. She was dressed in a gold lamé dress. Her brilliant soprano filled the great space. She was singing Monteverdi's *Lamento D'Arianna*. It was that very lament which had brought her and Strayhorn together. She'd become mute and was immensely fat, and Strayhorn was losing his mind and thinking of suicide. They'd found each other and saved each other. Theseus had deserted her, left her in Naxos.

> *Lasciate mi morire,*
> Ah, leave me here to perish.
> What comfort can you give,
> What solace offer?
> *O Te-seo, O Te-seo, mio. . . .*

She'd lost over a hundred pounds and looked great.

Bradovich curled up on his side and fell asleep. When he awoke, it was dark. The pain in his back was almost entirely gone. He could move with little discomfort. He felt strong, and was able to stand without dizziness. He could see the street lights and the headlights of cars flashing by.

The exit door as he approached it opened by itself. The seeing eye was still operative. It had not been a prison, after all.

Bradovich laughed, and walked out into the free air.

At the takeout counter of the Hunan Inn, two blocks south of his building on the boulevard, Bradovich bought containers of mu shu pork, thin pancakes, and rice. The Chinese clerk at the register always made the same joke with him. "You want flied lice, Blad?" "No, just boiled lice." They laughed, he paid his bill and left.

It had begun raining hailstones, and the streets glistened under the haloed night lights, the gutters overflowing so he had to leap across at the corners. When he jumped his back hurt badly.

When he entered the lobby, Slim stopped him to say, "Hey, you left so fast you forgot to close your door. Lucky Mrs. Kastner in 9D noticed it was ajar. I locked it for you. I looked in just to make sure nobody was inside. For a guy like you who's so damned careful, that's pretty dumb, ain't it?"

"You bet it is, Slim. Thanks."

Bradovich remembered making a specific effort to close his door that morning. They were getting at him in

38

every way they could. If they didn't get him one way, they would try another. The bastards would find he was no easy mark. He was a pro, goddammit, a pro.

At the mailbox, he found a letter from his daughter Laura. The envelope read DON'T BEND PLEASE so he knew it contained a photograph or two of Carlotta, his two-and-a-half-year-old granddaughter. He decided to wait until he was in his apartment, relaxing, so that he could gain full satisfaction and enjoyment from his grandchild's pictures and his daughter's letter, which he was certain would please him because she was a witty woman and wrote well.

Once in his flat, he walked through it slowly, giving it a close examination. Nothing seemed to be out of place. He sighed with relief, picked up the letter from Laura, then decided to wait before opening it, holding best things for last.

In his bedroom, under a strong lamp light, he removed The Authority's white linen shirt so he could examine himself in the long mirror which had been Val's delight. He saw he was expertly bandaged, and, using a hand mirror, examined his back as best he could. Skin above and below the bandage looked normal, so there was no inflammation as far as he could see. The skin under the bandage felt a bit tight, otherwise there was very little pain. Totally undressed, he washed himself thoroughly in the bathroom with a sponge, shaved—he had a date with Floriana that night—and returned to the bedroom to dress in clean clothes. The trousers he had worn earlier were covered with concrete dust from the floor of the old A & P. The shirt which The Authority had clothed him in was, he saw, from an expensive Fifth Avenue department store. He threw it, the trousers and everything else he had worn into the hamper in the bathroom.

In the dinette he set the table with chop sticks, took a bottle of beer from the refrigerator, and made a small dish of Coleman's mustard. The pork was still hot and he rolled a goodly portion in the pancakes, applied mustard, and began his meal. What was it they wanted of him? If only he could understand it. They say surveillance, then become brutal with him. Their word was certainly not to be honored. If they thought they could get him down that way, they were nuts.

Bradovich ate with relish. Guzzled his beer. He lived by himself, he could make as much noise as he fucking well pleased. "Yeah, to hell with them!" he said to the four walls. From the refrigerator he fetched a small can of pineapple chunks and finished off the meal.

The kitchen cleaned, his hands washed, Bradovich sat down in the easy chair in the living room and slit open the envelope from Laura and withdrew two photographs of Carlotta. The child stared at him with large black eyes—her father was a Mexican-American—her mouth wide open in a laugh. He could see the tips of several baby teeth. She was holding a big blue and red rubber ball in her hands and was obviously walking with it when the picture was snapped. The second photo revealed the child dressed in overalls looking back over her shoulder with a coquettish smile. Looks like Floriana, the little flirt. He laughed.

Bradovich gazed lovingly first at one photo, then the second, returned to the first. He ached to hold the child, to embrace her, to smell her sweet breath. He continued to shuffle the pictures in his hands, not getting enough of her. Sighing, he finally laid them on his lap and raised Laura's letter to read. She described the daily details of her life, made him laugh a few times with her description of Carlotta's antics. Before closing the letter, she again implored him to come for a visit, it wasn't fair to Carlotta, she should get to know her maternal grandfather, especially since she knew her paternal grandparents so well.

He just had to visit them. And Martin, too, and his son Alexander. Who knew how rough those bastards were going to get? Suppose it became a life and death struggle? Every pro knew if you got overconfident you could get yourself clobbered. Tomorrow he would go to the travel agency on the boulevard near Dominic's Italian Delight and have them book a round trip flight.

He read his watch, Floriana would be waiting. He could smell her already.

*F*loriana, nude, a red silk scarf tightly wound about her head, reclined on a pile of poufs and pillows of many and various colors. She was a woman of thirty-eight, with a ripe full body. Part Indian, part Negro, part Spanish, part French, her skin was light brown with a tinge of red highlighted by the lavender which dominated her pillows and poufs. Full-lipped, strong-jawed, her eyes were large and black, to enter them was to be easily lost. Now her eyes stared into the ceiling—one leg, a silk stocking rolled to mid-thigh, stretched full length, the other, bare, knee up, slightly canted.

"Manet's *Olympia* as done by Renoir," Bradovich said with a smile. He sat on a gilt chair placed so he could observe her in full glory.

"Tell me a story, Floriana."

"You're in bad trouble, aren't you, Wizard?"

"Let's forget that now."

"Golo will help you."

42

"He wants to marry you, why don't you?"

She laughed. "I enjoy my independence too much."

"He loves you, Flor. You ought to think about it."

"You're in deep trouble, the word's going out."

"Forget it, Floriana, damn it! Just tell me a story."

"What sort of story?"

"The most important story of your life."

"Why don't you tell me your story, Wiz?"

"Floriana, *por favor.*"

"*En su boca,* Bradovich."

"*Mi boca es su boca.*"

They laughed.

"Okay, fine, Wizard, you need diversion."

"*Muchas gracias, senorita.*"

"Once upon a time," and she smiled, "when I was a young girl, I loved my brother Raul, but he didn't love me. I never see—perhaps it is better to say, he never sees me, never comes to visit although he is now in *Norteamerica.*"

"Why, what happened between you and him?"

"Something of great importance."

"It must be if he never comes to see you. Who is older?"

"I am older—four years."

"Go ahead, Flor, it sounds like a good story."

"It's an old story, Wizard, but for me it is new."

"Old stories are the best."

"When we were small kids, we were always together. My father died early in life, and my mother had to go to work in the city. We lived in a small village outside the capital of our island. I had to take care of the house, watch my little brother. I was ten or eleven years old, and Raul was a small kid, maybe six or seven. He followed me all over the house and did everything I asked him to do, a good boy. At night, after my mama came home from work, we ate the food I

43

cooked, and then mother was so tired, she went to sleep.

"Raul and I slept in the same bed, a straw mat on the floor, and before we slept I told him every night a story. Then I hugged him tight and he went to sleep. Afterwards, I cleaned up the dishes from supper, then patched clothes and the like. One time, he was only seven or eight, when I hugged him his thing got hard. He was surprised—and me too. I told him it was a sin, but he didn't understand what I was talking about. He went to sleep, a sweet smile on his beautiful face. He is even more beautiful than me, and I am very beautiful, aren't I, Wizard?"

"Every inch of you, Flor."

Floriana resumed. "When I was twelve years old I began to menstruate, my breasts began to grow, and my hair here began to sprout."

"And a glorious bush it is, Flor."

"Yes. My mama said it was time for me to sleep alone. She put me in the same room with her and Raul slept in the other room by himself. Everything else was the same. My little brother was with me all day, except now we went to the church to be taught how to read and write by the nuns. Life went on not too badly because my mama worked so we had enough for food."

"Lucky, I'd say."

"Yes, lucky. One day I was in the back of the house taking a shower from the pail, and Raul came in, too. He hugged me and I let him. His thing became hard and I let him put it between my thighs—not in, just between near where it is soft and tender. He was still too young to come, but he rubbed and rubbed. I liked it, too. He rubbed and rubbed, and, oh, man, I hit it real good, and had what looked like a crazy convulsion to Raul and he got scared and ran away. I felt guilty, I knew it was a sin, forbidden, especially with a brother. But I told Raul it was all right, it gave me great

44

pleasure, and soon we became crazy for it. He came to me every afternoon and just rubbed and rubbed with his little hard thing until I came.

"Other boys asked me to go with them to dance in the fiesta, but I said no. Soon Raul was a man, twelve years old, and I am sixteen, a wo-man. I loved Raul, only him. One day, my mother went to visit her sister Lizeta in another village, and she stayed overnight, and I was in bed reading a book about the saints and Raul came in to me, undressed, and he had a big erection, and he jumped into bed and right away wanted to put it in where it is tender. I couldn't say no, and I let him. He had a long hard fat one for a little boy. It hurt plenty, but I loved it. He came quick, but he was ready to put it in again soon. That long hard fat thing, it moved very fast. It was great. It was a terrible sin, but I didn't care, I liked it more because it was a sin. I did what was forbidden. Soon Raul and me, we did it all the time when mama wasn't home. We didn't have to read how-to books, we learned by ourselves. *Mi boca es su boca,* I told Raul, and he understood, and soon we did it that way, too, and like dogs. Oh, man, oh, man, we had it good.

"One night when we ate supper, mama asked me why I did not go with the boys in the village and I told her I didn't like them, I was happy to be at home with her and Raul. She looked a long time into my eyes but said nothing. We went on this way until Raul was fifteen years old, and he began to go with other girls, and when I asked him why, he laughed and told me I was an old wo-man now. I wanted to kill him. Now when he came to me with his long hard fat thing, I said no, go to your girl friends. Then one of the boys in the village, the son of mama's good friend Teresa, began to pay me attention and now it drove Raul crazy. He was like a husband, he could do what he liked, but his wife, oh, *mamita!* He said if I didn't stop going with Tomás he would kill him.

45

He had a knife and always sharpened it on a stone in front of me. I laughed at him and continued to see Tomás every day. Soon I began to feel free of my brother and Tomás asked me to marry him. I said yes and my mother and Teresa and everybody in the village were very happy for us. For the first time I confessed to the *padre* about Raul and me and he said now that I am marrying Tomás I will become a good woman.

"One day soon Tomás was found in the woods near the village stabbed to death. No one knew who did it except me. The priest asked me and I told him I didn't know. He didn't believe me, but what could he do? Afterwards, when Raul tried to do it with me, I said no, you are a murderer. He laughed, he had no feeling about what he had done. And when my mother wasn't home he walked around the house naked, his long fat thing hard and red and finally I couldn't stand it, I said yes, yes, give it to me. Now we were like wild animals when we did it, bit, hurt each other, never had enough. We ate each other's flesh. And I wanted his hard thing in my mouth, between my legs, in my behind, between my breasts, and he was like a fountain that spurted day and night.

"Then, of course, it happened, my month didn't come. I didn't hesitate, Wizard, I can tell you that. Late at night I went to a certain old wo-man in my village, and she gave me a knitting needle and told me what to do with it. My mother found me bleeding to death, and she screamed and cried, but she nursed me, and I began to get better. When I was all better again, mama became sick, lost weight, got skinnier and skinnier, and soon my poor mama died. I felt it was my sin, Raul's and mine, which killed her. But I still loved him like a mad wo-man. But he changed. He wasn't interested in me after my mother died, he wrinkled his nose like I stank from garbage, and soon he left for the big city, went into the army, and I didn't see him for a long time. When I did see him, he couldn't look at me, he acted as if it was all my fault that we sinned. If it was up to me, I would sin with him the rest of my life. I love him."

Floriana was silent now, and Bradovich did not interrupt that silence. At last, Floriana said, "That's my story, Wizard. Do you like it?"

"I've heard it before, Floriana. I've heard it a thousand times. It happened to Val, and I have a beautiful sister I both loved and hated, so I understand."

"Did you do it with her?"

Bradovich didn't answer, but remained silent.

Floriana broke the quiet. "Now, please."

They were sitting at a small round gold pine table in Floriana's bedroom, drinking tea. She was wearing a cardinal red silk kimona, and her heavy black hair was loose about her shoulders.

"Is that why you won't marry Golo?" Bradovich continued. "He's a good guy, Flor, he'll treat you fair and square, decently."

"Yes, I know. No, no, I will never belong to another man. I do what's forbidden, that way I'm free. My own woman."

"You're just kidding yourself."

The tea was very hot, and Bradovich rippled it with his breath to cool it.

"If a person lets it, Wizard, guilt can kill him or her. You understand that, don't you?"

"You talking about me or yourself?"

He saw her put on her wise owl face, lips pressed tight, black intelligent eyes wide open. "In general," she said softly. "Just in general, Bradovich."

There was no sense denying it, so he said, "I'm under surveillance, that's all."

She stood, as he did, and she kissed him. "If you need help, there's Golo, of course. Besides, I know important men, too."

"Thanks, Flor. You're kind."

He embraced her one last time, stepped back, bowed, and headed for the door. Consuela, Floriana's beautiful young maid, let him out. A few moments later, as he made the turn a half flight up, he heard distinctive and familiar footsteps along Floriana's corridor. Looking down, he saw Golo the Gimp hopping along on his little feet.

*H*is door was ajar. He'd had enough for one day. "I'm going to kill you!" he shouted, certain whoever it was was secreted somewhere in the apartment. He stepped inside and locked the door behind him. Began a search for the intruder. Closets, crannies, under the beds. He found no one. Nothing had been touched so far as he could see. In the kitchen sink he saw a cup with coffee dregs. He hadn't had coffee for dinner; he'd had a bottle of beer. Someone had opened his door, entered the apartment, drunk a cup of coffee, placed it in the sink, departed leaving the door open. A signature and an exclamation point. Calmly, in perfect control, Bradovich picked up the cup and smashed it against the wall. Then, as if transfixed, he watched as the coffee dregs slowly slid down the wall in thin brown black lines.

The phone's ring startled him, and he unconsciously backed away from it as if it were a striking snake. It rang and rang in the quiet apartment. At last, he moved toward it and raised its hook. He exhaled with relief. It was the familiar voice of the caretaker of his studio building downtown. The door to his studio was found open, no one was in there, nothing seemed to have been touched, should he call the cops?

"No, no, I'll come down myself."

There was no sense in calling the cops, it was The Authority. Whatever or whoever they were, they were supra-official. Bradovich stood for a moment paralyzed, his mind a blank. He shook his head free from what held it. He was scared. Admit it, you fool. It can happen to the best of us.

On the run he grabbed up a coat from the closet, made double certain to lock the door, and ran down the nine flights rather than wait for the old elevator.

The night was cold and black gray. A taxi stood at the curb, discharging two passengers, and he caught hold of its rear door before it slammed shut. It was Beatrice Holden and a male companion, still holding the playbills in their hands. He nodded and jumped in. As the cab left the gutter with a lurch, Bradovich saw leaning against a corner of the building a tall stonefaced woman in a gray overcoat and wearing a black fedora. She was speaking into a cordless phone. Reckless, Wizard waved a debonair so long.

What during daylight hours would have taken a good hour at 11:45 at night took less than fifteen minutes. The caretaker was there waiting for him. "Didja forget to lock her up before ya left, Brad?"

"No, dammit! I bolted it twice."

Nothing had been touched, neither his tools, nor old heads he had done years before, none of the new blocks of marble awaiting his chisels. As he gazed about from the middle of the floor, he saw in a darkened corner a wood head he had once done. An attempt at a self-portrait and never finished, the face was just emerging from the cedar six by six he had used, its mouth agape, eyes but no nose, and now it hung from the high ceiling with a tight noose around its clumsily formed neck. In this generally brightly lit and very whitewashed room, the figure hanging in the darkened corner appeared so real to him that he just barely repressed a cry of pain. "No! No!" Then with pure bravado, he found a large square of the white cardboard he used for sketching, printed on it in large black letters, THE AUTHORITY, and tacked it to the unformed chest of the figure.

Bradovich had a cot in his studio, and since it was late decided to sleep there as he had so many times in the past when working on a piece so totally engrossed him he'd forgotten time. After the children had grown, Val would come to stay those nights with him, bringing with her something to eat, and they'd unfold the thick foam slab on the floor to sleep on and so many times to make love. "You're such a damned tease, Wiz. Hurry up, you big hunk of loveliness."

Or talk.

"Don't die before me, Wizard, I just couldn't bear it if you did."

"Don't be an idiot, I've got ten years headstart on you. I have to die before you, and I will will it so. Anyway, you're the marrying kind, you'll get another guy fast. Every painter and sculptor in town will be sniffing at your cunt."

She laughed. "Will you be jealous up in heaven?"

"Or hell. No. You know I'm randy, I'll enjoy watching."

"Ugh. I'd hate it watching, and I'd die if I ever walked in on you screwing another woman."

"I'm as pure as the driven snow, what the hell ever that means."

"You're putting less on your stone heads than Brancusi ever did. You can't go further, Bradovich, you won't

51

have anything left. You're driving your gallery crazy, and the critics are beside themselves with not knowing what to say."

"And what else is there for me to say? All right, you push me to the wall. What else is there for the human shit-ass race to say? Didn't our lovely, rational erect-walking species say it all in the Holocaust and the Gulag? Idi Amin and Pol Pot put a blood red exclamation point to it. There's only one thing left, isn't there? The Big A or the Giant H."

"It will never happen, Wizard, I feel it in my bones."

"Optimist. You only see the decency in people."

"It's a better way to live than your way, raging against the world."

"I'm an artist, I'm trained to see deeper."

"Oh, come on, Mister Bradovich, Arteest, you going to start throwing that shit, especially to a poor little hunched-back librarian like me?"

He'd laughed, and she'd joined him, not hunchbacked at all, straight as an arrow. That arteest bull shit. Those political nitwits who made large grandiose statements and thought they could get away with it because they were artists. For God's sake, most of them didn't know how to spell, let alone how to read.

What was he thinking about? He had his head in a noose. It was there in that corner. All the bravado in the world couldn't conceal it now. They were after him, and they meant business. Like everyone else caught in the same noose, he asked why me? It seemed to be arbitrary, helter skelter.

He tried to sleep, but sleep was like a greased eel, uncapturable, elusive.

Val had been an optimist, saw the good side before even venturing an eye to the bad. "Basically, most people are good," she said.

"Basically, most people stink, are unwashed, too busy grubbing in the mud for themselves," Bradovich answered.

Now he complained to the night, "It's the good ones who get taken to the cleaners first. Bang!"

He missed Val, missed having a woman to talk with. On her dying bed she'd said, "It will be hard for you. Take a good look at Beatrice, she likes you, and she'll be good for you."

"Oh, come on, Val, cut the crap. Besides, she's not as pretty as you."

Even dying, she got mad. "Is that what you're going to miss, my elegant nostrils?"

He'd cried.

Now he turned and squirmed the night away, the wound in his back stinging at every turn. He'd forgotten all about it when he was with Floriana. But he liked the pain, it was good for him. There had been times when he'd give one of his leather belts to Val and ask her to whip him.

"Are you crazy, Wizard? Cut it out."

Once, tired of his asking, really angry with him, she had taken the belt and whipped his naked back until he asked her to stop, he'd had enough. Then she'd looked at him, those wise eyes of hers black with rage. "Did you get your kicks, you dumb bastard?"

"You don't understand, Val, it wasn't for kicks. That was furthest from my mind."

"You haven't been screwing somebody else lately, have you?"

"Of course not."

She now took the time to look into his face, and slowly her anger ebbed away. They stared eye to eye. Then she said, "Oh, my God!" She was quick-witted, sharp, now understood.

Optimist though she was, she wasn't sentimental,

and knew enough not to break down and take him in her arms in pity. He would have hated that. She herself always said, one way or another, sentimentality was false sentiment just as nostalgia was false history, a lie for those for whom uncomfortable truth was always to be avoided.

Leave it to the Jews to have the right word for Val. She was a *mensch*.

Bradovich slept fitfully. When he awoke the morning sun had begun to creep in through the barred windows and the skylights. In the corner the hanging head with its agape mouth and staring eyes swung tremulously under a single ray of cold morning light. He'd leave it hanging like that with THE AUTHORITY sign tacked on just in case they returned. Let them see they couldn't intimidate him.

He dressed, washed in the small bathroom at the back of the studio. Chose a six foot length of cedar six by six and in addition to barring the entrance to his studio with steel bars banged in the cedar stud with large spikes. As he handed the heavy hammer to the caretaker who had helped him, he said, "If you see anything, call the cops. If you want to shoot them, go ahead, be my guest, I'll take full responsibility. I'm going to the coast to visit my kids, so if anything happens call Golo the Gimp, you have his number. Thanks, bud."

"Okay, Brad. Have a good trip, and regards to Marty and little Laura."

"Sure thing."

The narrow cobbled street was, as usual this early in the morning, clogged with numerous trucks, backing up, inching forward, their diesel and gas fumes permeating the air, yet hardly noticed by those who worked in the area. In no particular hurry, Bradovich strolled slowly down the street toward a subway line which would take him to the east side of the park, from which a twenty minute walk would bring him to the boulevard and his apartment building.

As he approached the subway kiosk, he saw that a brawl was in progress. Three familiar looking square-shouldered men, wearing identical black trousers and black woolen turtleneck sweaters, were pummeling a man brutally. As they did a crowd gathered and no one interfered. Someone did say the police had been called. Bradovich pushed his hulk into the midst of the melee, shouting, "Cut it out, you guys!" One of the attackers stopped beating the victim long enough to turn to take on Bradovich. Wizard was glad to accommodate him and lashed out with a powerful jab that knocked the hoodlum on his butt. Before Bradovich could turn to the next roughneck, someone, probably a confederate, yelled, "Here come the cops!" The thugs stopped pounding the man and seemed to disappear into the cobblestones underfoot, so quickly were they gone—including the man Bradovich had knocked to the ground. The victim who'd been taking an awful beating, his face a mask of blood, turned to Bradovich and said, "You ought to mind your damned business," and then disappeared down the street.

When the police cruiser arrived, the crowd had already dispersed. Bradovich remained and when he approached the police car, hoping to tell them what had occurred, one of the cops said, "Forget it, mister, be glad it wasn't you."

Intemperately, Bradovich raised his huge fist, ready to punch the cop in the face, but the cruiser jerked forward and continued down the street now suddenly empty of pedestrians who'd entered their workplaces. Once again it was just a rundown desolate street, the huge garage doors of the warehouses hidden by the trucks backed up to them waiting to discharge or load.

Bradovich slumped into the underground, bought a tabloid to read for his journey home. The platform was forbidding in its dimness and smelled of dust and urine. The train, when it finally jolted into the station, was still crowded with rush-hour passengers, pale, haggard and fatigued even before the day's work began. He was able to elbow his immense body into a corner at the end of the car and to look at the tabloid filled with pictures of the players in the latest melodrama, today starring a demented woman who had murdered six of her eight children. On page 10, there were airphotos of another war starting up in Africa.

Bradovich stood before the door of the Cosmic Travel Agency at 10 a.m. It was closed. A sign read WILL OPEN AT 2 P.M. When he returned a few minutes after 2, the sign read AT 3 P.M. At 3 p.m. the sign said the agency would not reopen until the following morning at 10 a.m. He decided to go to another travel agency on the boulevard not far from his building. When Atlas Travel was found, it also was closed, its sign reading that it would reopen in the spring.

Now he decided to phone an airline which serviced the west coast directly. He kept getting a busy signal. He dialed all hours of the day and night. After innumerable attempts he finally got through and was told to hold until a booking clerk became available. They bored him with non-music music. He held on for forty-five minutes, then gave up in disgust.

The following morning he retraced his steps to the

Cosmic Travel Agency. Again he found it closed, with a sign indicating it would be open at a later hour. And yet again he found it closed, this time the sign reading it would be open the following week. He returned to his building. Before entering, he encountered Beatrice Holden, her red gray hair a heavy cloud above her head. To his astonishment, she clasped his head in her two hands—cool, long fingers—stared deeply into his eyes, kissed him full on the mouth, then hurried away, before he could say a word, her long slender legs flowing into a thoroughbred's ass. He followed her with wide puzzled eyes and agape mouth. What the hell's going on? Then he smiled. What a sweet, gentle kiss.

In his apartment, he dialed another airline, busy signal after busy signal. At last he got through. And again non-music on hold. This time he almost broke the telephone slamming it down after almost fifty minutes of the non-music.

As he sat at the dinette table, breathing heavily, mouth dry, it hit him that The Authority had made a determination that he was not to leave the confines of the city. He was on his own recognizance, it was true, but with limits.

Besides he had become aware that wherever he'd gone, a shadow had tailed him, or, was it, a tail had shadowed him? It became impossible to chase every damned one of them, so he'd decided to ignore them. Man or woman, they wore black suits, black fedoras, and had faces etched in stone. African faces. Asiatic faces. European faces. Waspish faces. On occasion it would be a pair of them, twins. He had begun, he noticed himself, to accept them as one accepts one's own shadow. Of course, if one should approach too close, he would wallop the man and/or woman, but it also became apparent to him that they had learned to be wary, he was no easy mark. Sometimes he gave them the finger. A few times he yelled at them, "Up yours, you bastards." Most of the times, he merely muttered his usual "Fuck 'em!"

Bradovich was a stubborn man. Had been all his life. Had driven his parents, sister, relatives and friends crazy with his stubbornness. Enemies, of course, had to back off or he would destroy them. "A tough cocksucker," was the universal opinion of friend and foe alike. He had decided that The Authority would not defeat him. Still, as of now, it was senseless to beat his head against a stone wall. He would cool it for a time. See what Golo would come up with.

He dialed his daughter on the west coast and with a sigh of relief got through. "Pops! How good to hear from you. When you coming?"

He told her he was held up by a snafu with a misdirected shipment of one of his polished marble heads—a simple lie seemed easier than the complex truth—but he would definitely be coming as soon as he straightened it out.

She seemed to believe him, and then went on to tell him about Carlotta and her husband and her life as a professor of philosophy, her disgust with academic politics. "Ideology is becoming more important than intelligence, and ideologues than wise men and women."

He agreed with her, told her his show in Rome had been well received and that Mitsubishi had paid a big price for one of his painted heads. "I'm doing better now than I ever did. But what good is it now that Val's gone?"

"I miss her, too, Pop. Awfully."

They were both quiet, then he said, "Call Marty and tell him why I can't come now."

"Okay, Pops. Be well."

"Give Carlotta and Gonzalez a big hug and kiss for me. Marty and his family, too."

When Bradovich hung up, he had tears in his eyes. He was not afraid of tears. He was afraid of sadness.

59

At a renowned and mammoth toy store he bought several gifts for Carlotta and Alexander and had the establishment air express the toys to the west coast.

Of course, on his way home he was shadowed.

*T*here seemed to be more of them, and they seemed to be getting closer. Tall square-shouldered men and women, in black suits and black fedoras, sometimes wearing black turtlenecks, other times white shirts and black ties. Stony-faced operatives in gray overcoats. Their skin color varied.

Bradovich's jaw jutted more stubbornly, his large hands fisted, knuckles white. He took to staying up later than was his custom. Movies, television, his few friends aside from Floriana and Golo made him impatient. Many times he began to dial Beatrice Holden's phone number but stopped midway, why bother her? She was a happy woman, why make her sad? He did see Floriana on occasion, but Golo was on tour in Europe. Wizard was anxious for Golo to return. He didn't impose on Floriana's offer to help with the important and powerful men she knew. If a man did a favor for her, he demanded payment under the arch. Bradovich was not about to encourage that.

His own crowd of artist friends—and most of his friends were artists (as a group they were very sectarian, being too high and mighty for the rest of the world, Bradovich no exception)—all seemed to be aware that he was in some sort of trouble and with an obviousness that he found patronizing, they avoided mentioning it. The exceptions, those very few who made an effort to ask him about it, he became angry with, telling them to mind their fucking business. "I'm all right, leave me alone."

First, he just stayed up late, reading or looking obsessively at colored prints of Brancusi's work. "Best goddamned sculptor of the twentieth century. Even better than Giacometti—but, what the hell, that's like comparing oranges and apples, it depends on what you like better." Then, feeling penned in, ignoring the operatives, he would walk the boulevard, even venturing into the park at night. He never worried about assailants, as big a man as he was whose step was alive and forceful.

One night, long past midnight, he decided to cross the park to the east side of the city and back. So far as he could see, no one tailed him, and he encountered no one. Only those foolhardy souls who courted danger ever entered the park at night, and perhaps that included his shadows, too. He laughed. The birds, even the squirrels were asleep. If the city rats were about, he saw none. He whistled, he sang. Felt absolutely free for once. In the distance, he heard an occasional car horn. To his left ahead he saw the dark Egyptian obelisk giving the night the Roman salute, a very Bradovichian symbol. Ahead, a dim light could be seen in the tunnel leading to the great museum on the avenue. Totally self-absorbed he entered the tunnel and as he did was abruptly accosted by two youths, each with a Mohawk haircut and a face painted with sulphuric yellow that gleamed in the dark. One held a long butcher knife, the other a meat cleaver. "Take it easy, kids," he said.

In the center of the tunnel a debris fire was bursting with flames, popping little explosions. The Meat Cleaver and the Long Butcher Knife hemmed him in. As his eyes focussed, he saw the fire was ringed by many young men and women with Mohawk haircuts and painted faces, all dressed in overalls and shirts of varied primary colors, a uniform of sorts. He knew that to betray fear was dangerous. He made an effort, and succeeded, to stand impassively, to betray nothing. The Meat Cleaver ordered him to come closer to the fire, where those sitting could better observe him. He did as he was asked.

"Does he suit you, ladies and gentlemen?" the Meat Cleaver asked, and after a few moments during which they examined him closely, those sitting nodded in affirmation.

"No challenges?"

There followed a unified shaking of heads in the negative.

Politely, the Meat Cleaver explained to him that a trial was in progress, one of their number had committed a flagrant crime against her peers, and would he act as judge in good conscience?

"Do I have to?" Wizard asked, controlling his voice as best he could. He believed he had managed it well, for he felt that if he gave any sign of panic it would be turned against him.

"Look, Mr. Wizard"—the youth seemed to know him—"what else have you got going for you tonight?" He raised the cleaver over his head, not threateningly, but there it was, glistening in the fire's light.

Deciding quickly, Bradovich said, "Sure, why not?" He hoped to get it over with as quickly as possible.

The Accused, a young woman whose face was painted with some sort of white cosmetic and whose head was shaved leaving only an Apache topknot at the back was wearing a long tight-fitting sequined gown cut so low her small naked breasts protruded. Against her will she was pushed close to the fire by two armed women guards (they carried Saturday night specials whose short barrels they pressed against the young woman's sides). Both guards were dressed similarly, with like hairdos as the Accused. Their faces, however, were painted a luminous red.

A sharp white light was now focussed on two people sitting on thronelike seats, a man and a woman, both dressed in royal ermine and sable, crowned and sceptred. Brilliants gleamed on their fingers. Their faces were also painted an opaque white, but both heads were totally clean-shaven under diamond-studded crowns.

The Queen, addressing the Accused, said, "You did, didn't you, my dear?"

"I did what, your dear?" asked the Accused with an arrogance that Bradovich thought foolish for someone in her precarious position.

Speaking with a slow stammer, the King said, "You know very well what it is you did, don't you, little miss?"

"I did what I did, I admit that," the young woman on the dock said forthrightly. "I have a perfect right to do as I do. You have to prove otherwise. Every person has a right to do as he or she does. I did."

The diction of these strange people astonished Bradovich not a little. They were not, as he first thought, illiterate street children. But what was it that she had done, he wondered. Around the fire, the sitters mumbled among themselves, no words distinct, to him at any rate. He assumed they were taking sides, because there was angry gesturing and pointing of fingers at one or another of them, but which side they took he couldn't fathom.

The royal couple and the Accused seemed to have

arrived at an impasse. They were silent. The Accused stood defiantly. The Queen stared stonily. The King seemed to have difficulty focussing his eyes.

The Meat Cleaver, a bailiff of sorts, said in a very loud voice, "Mr. B., what do you think about what has taken place here? Be careful, however, of what you say." He spoke courteously, his weapon at attention on his shoulder.

Bradovich stared intently at the Accused. She was standing fully erect, shoulders thrown back, small pointed breasts aquiver, chin high, her eyes flashing in the firelight, her lips sullen. A stubborn kid. A model for all those who stand at the dock. No fifth amendment for her. She did what she did. Had done what she'd done. Would do as she would. Notwithstanding.

Then Bradovich turned his gaze to the Queen and King. They sat back in their thrones, wholly relaxed, looking at him with an aristocratic contempt derived from royal arrogance. It seemed hardly to matter to them how he decided.

Now he observed those sitting around the fire. They were murmuring among themselves, deciding for themselves, no matter what he had to say on the matter. The Accused, after all, was their peer.

The Meat Cleaver shouted, "Oyez, oyez, the Court is now in session, the Honorable Wizard presiding."

Everyone but the King and Queen stood, bowed to the Court, then reseated themselves.

Her proud breasts quivering with every breath, a mere sheen of perspiration on her painted opaque white face, the young Accused stood stubbornly impassive.

The bright light soloed in on Bradovich. He had, he believed, to make a life and death decision. For the Accused. Even, perhaps, for himself.

His throat was dry, and he raised a pitcher that rested

on a little table near where he stood and poured water into a glass tumbler there for that purpose. Drinking slowly, he closed his eyes and gave thought to the question before him. He wondered whether he was himself really in danger, put the question aside quickly as unfair to the Accused just as the Meat Cleaver proclaimed loudly, "Whether or not, whenever or not, did she or did she not? Your judgment is awaited, sir." The man's weapon was raised to what could be called a threatening position.

Bradovich looked about again at the arrogant faces of the royal couple, at the arrogant cruel smugness of the Accused's peers, then clearing his throat with another sip of water, spoke ringingly. "The Accused has admitted that she did what she did. Her Accusers say she knows what she did and ask should she have done what she did? She rejoins, however, that she did what she did as a natural right. For the thousands and thousands of years that men and women have roamed the earth it has never been determined beyond a reasonable doubt what is or is not a natural right. I therefore rule: the Accused in all justice cannot be adjudged. That is my verdict, pending appeal as to the correctness of the law hereunder. So ordered."

Having given this decision which was received with a gasp by all those present—the Accused yielded up a small dignified smile—Bradovich spun quickly around and left the tunnel on the run, expecting at any moment to feel the meat cleaver rending skin and flesh between his shoulders. At first he thought he heard footsteps dogging him, but soon realized no one had given chase. Not even a shadow on his tail.

Soon, terribly fatigued, he was in his apartment undressing and preparing for bed. Before clicking the light off, he caught a glimpse of his face in the bureau mirror. It was as opaquely white as that of the stubborn Accused. What a marvelous young woman.

Just before he fell asleep, the legal term *nolo contendere* flashed into his brain. No contest. No! No! that's already a concession. He would fight them to the end. No concessions!

*W*hen Bradovich awoke several hours later, he could see that the sash under the open window was wet from melted snow. Sitting up, he saw it was snowing heavily, the flakes like clumsy white moths. As he approached the window to close it, he saw Mrs. Kastner, his courtyard neighbor, cleaning her window sash of snow with a whiskbroom. She saw him standing naked as she had many times before—she couldn't see below his bellybutton—and stuck her tongue out at him, and he grinned. She returned the grin, her tongue lolling in her mouth like a thick red fish. With her dyed blond hair askew (as it always was) she looked like a gargoyle. Before he closed the window, she stuck her head out and called, "Before you go in to pee, why don't you bring that stiff cock over to me?" She was only kidding, they were old neighbors and understood each other.

"It's the wrong time of month," he called back, and they both laughed.

"Beatrice is dying for it, why don't you give her a break, dummy?"

He waved her away and shut the window with a bang. Everyone, but everyone was enlisted on a Beatrice Holden crusade to capture Bradovich as if he were the Holy Land. What good would it do her now? Like the Holy Land itself he was being claimed by others, laid seige to, a war, perhaps, to the finish. What was that business last night, had that trial been real? That tough kid.

His window closed, he fetched the mop from the broom closet in order to dry the floor. Looking out the window before washing and dressing, he couldn't help thinking back to his youth. How joyful had been the days when it snowed, going with his buddies to a park nearby, Lincoln as he remembered, sledding furiously down the short but precipitous hill, then running like crazy back up, and racing down again. Hundreds of bellyflops—nose running, toes tingling, his mother having hot chicken soup on the stove waiting for him.

Innocence. At what point did the innocence slip away and the corruption begin? It was like what happens to a perfect block of marble when you begin to pitch with a crude chisel, suddenly sneeze, and there it is, a destroyed piece of stone. No, it was more like erosion by time. Unlike the proverbial fall as a result of serpentine seduction, your innocence was eroded by an accretion of evil.

Was that a paradox? Should have listened more to the philosophy professor than to the football coach. He would have to ask Laura who taught philosophy at a state university. Erosion by accretion. Was that an oxymoron? To say a man is innocent is itself an oxymoron. Man innocent? No sooner are we born than we begin the accretion-erosion process. Was physical erosion itself a loss of innocence?

Since the beginning of time, men and women have polluted the earth, it wasn't after all a condition of modern times alone. In order to live, we destroy. We kill to live, don't we? Always have. Will it ever end? You have to be a goddamn idiot to think it will. Long live the idiots!

Bradovich stood shivering before the window, closed though it was. The wound in his back suddenly stung. He had removed the bandage several days before because it had stopped hurting and the mirror had shown all the redness gone, the scar imperceptible. He examined it again, using the hand mirror as he stood before the full length one. No, it wasn't inflamed again. Just a momentary sting. To remind him they were even in his room. *Fuck 'em!* It was all better. He was one of those lucky people who healed quickly. Still, pain wasn't all bad, it proved you were still alive. It was also balm for your sins. A sin can't be all that evil if you suffer for it, can it?

He showered and shaved, instead of dressing slipped into the Viyella plaid robe Val had bought him when he hit fifty-five, then went to the living room to stare out at the boulevard. The snow was falling heavily, the cityscape like a backdrop seen through a white mesh scrim. Tall two-dimensional buildings, stick figures stomping through the drifts forming on streets empty of most vehicles. Cars parked at the curbs already looked like large humps of snow.

Bradovich stood staring out the window at this wavering backdrop for many minutes until, hypnotized, he found it impossible to keep his eyes open, and he returned to bed, pulled the covers over his head. When would they come again? As long as the stonefaces kept their distance, he could say fuck 'em. But they were not fooling around, they meant business. Surveillance. Then the grand jury. Then the trial. For what? What the hell had he done? And that mock trial last night? Had that been staged just for him?

From across the courtyard at Mrs. Kastner's open window—she was an open air freak—the words of a song slipped out in her cracked voice.

> I held firm
> Like a rock.
> I was strong as could be.
> Like a rock.
> I was something to see.
> Like a rock.
> I still believed in my dreams. . . .

She sang, she sang, she sang. . . .

*B*radovich screamed as the blanket was ripped from his body and he lay naked and vulnerable under the bright ceiling light of his bedroom. The shades at the windows were down. He tried to sit up, but found his ankles and arms held down by four men, each at a separate limb. "You nogood bastards!" he cried out. Two square-shouldered men, the twins of the first day, stood over him. One held a truncheon.

Bradovich struggled and one arm broke loose, but they grabbed him and secured him again.

The Truncheon angrily said, "He's too damned stubborn, this one. I'm going to really give it to him."

"You're not to beat him," the second said. "There has been no decision as yet."

"He's pleading *nolo contendere*," the Truncheon said, rubbing the ball of his hand brutally into Bradovich's chest. "So what makes the difference?"

"I withdrew the *nolo*, you son-of-a-bitch," Bradovich yelled, trying desperately to free his hands.

"Too late, Mr. B., just a bit too late," sneered the Truncheon, slapping him hard on the face with a free hand.

Cunningly, Wizard said, "Okay, go ahead, beat me, I deserve it."

The twins paid him no heed, but continued to argue the niceties of The Authority's law.

"The code states that until a final decision is reached there's to be no extraordinary physical punishment, only harassment."

"That knife in my back, is that what you call harassment?"

They ignored him. As far as they were concerned, he wasn't even there, though he kept hurling curses at them.

"But when an alleged defendant," the Truncheon began, walloping Bradovich on one of his knee caps, "is as obstinate—"

"Mr. B. isn't a defendant as yet, he is merely being investigated. There is still plenty of time to go."

Bradovich was grateful for the crumb, and kept his mouth clamped shut, though the clout on his kneecap had hurt terribly.

"He pleaded no contest," the Truncheon said, seemingly obsessed by the idea. "That's an indication that he is preparing to concede. Isn't that so, Wizard, old boy?" he said, this time poking the truncheon sharply into Bradovich's side, so that he gasped.

After catching his breath, Wizard said angrily, "I withdrew my plea of no contest almost immediately after I made it. I'm fighting this to the very end."

The Truncheon heard this remark and sneered. "You're not kidding, bud, the end."

"The law's the law," the legalistic twin continued, "ass or no ass."

"Dickens would be glad to hear that," Bradovich interjected. He was in pain but was damned if he would show it to these hoods.

"And it says physical punishment is to be withheld until The Authority makes its final determination, that is, following all appeals. We are merely here to remind Mr. B. that he's still under surveillance, pending further review, and then trial." Turning to Bradovich, he said, "You understand that, don't you?"

Sensing that The Authority's operatives were preparing to release him and leave, Wizard said, "Yes, of course." He waited patiently now.

"You are on your own recognizance as before, but must confine yourself to the city's limits. There will be no violence perpetrated upon you."

Whereupon, the Truncheon swung his weapon with great force against the side of Wizard's head. Fortunately, Bradovich had seen the blow coming and had moved his head swiftly enough to avoid receiving the blow with full force, so that when his limbs were released, he was able to recover his senses more quickly than otherwise would have been the case. He leaped from the bed and struck the Truncheon on the back of the head with a powerful right cross. As the man went down to his knees, Bradovich axed him on the back of the neck, and the man fell senseless to the floor. Before Bradovich could turn on the others, one of them kneed him in the groin, and he had to step back against the bed. Now another of the operatives knocked Bradovich cold with a billy club to the temple and he fell.

"This will have to be reported," one said.

"Yes," replied another.

They raised their fallen colleague from the floor, and

with quick steps moved toward the outer door of the apartment. Bradovich, recovered from the blows to some extent, went after them. By the time he made it to the outside corridor, they were gone. He could hear their footsteps down the stairs. He lifted the house phone and rang Slim. But there was no response. The guy was never there when you needed him.

Drained of adrenalin, his head and body aching, Bradovich could barely make it back to his bed. Before he lay on it, he caught sight of himself in the long mirror. There was a long bloody welt across his abdomen, and the side of his head was beet red.

On his back, staring at the ceiling, he felt more tired and beaten than ever before in his life, even after the toughest of ballgames. This was becoming more and more serious. Perhaps deadly. As one spoke about the niceties of the law, another operative whipped him. They manufactured their own law as they proceeded in their harassment. Golo the Gimp was due back that very night. Better call him later and make a date to see him tomorrow.

He needed some fresh air. He crawled from the bed to the window, raised the blind, then the window. Mrs. Kastner was leaning out of hers into the narrow courtyard. No doubt looking for salacious tidbits with which to enliven her husband. She always complained he never gave her enough, had to be roused with porn. Seeing him, she smiled, began to crank out a song in her raspy voice.

> You will be the clothes pin,
> I will be the clothes line.
> We will hang out together, ta ra ta ra. . . .

*I*t was a gray day—gray the sky, gray the faces of those who ventured out into the streets, gray the snow piled high along the curbs.

Dressed in gray, his skin gray because of troubled sleep, his abdomen throbbing, the welt still red, his body an ache, his head ringing, Bradovich slowly walked on the slippery sidewalk. He was on his way to a bus stop, headed downtown to the office of Golo the Gimp.

At the corner, the light was against him, and he leaned tiredly against a post to await the light's changing. It was his hope that Golo would be able to resolve his problem, put a halt to what was turning out to be a little more than plain harassment, surveillance they called it, though his two shadows back there near the discount pharmacy countered all hope. The light turned to Walk, but before he could step off the curb, Beatrice Holden approached him. She wore tan boots, a long tan loden, and a tan hood. Her pointy nose stuck out cold red, and he thought she looked like a pretty rabbit. A tall one, a long drink of water.

"Hello, Beatrice. I never had a chance to return that sweet kiss."

"What are you waiting for, then, Bradovich?"

He leaned toward her and kissed her lightly on each cheek.

"Thanks so much, Bradovich. It seems my lips aren't good enough for you."

"We're not eighteen-year-old kids kissing on street corners, for God's sake."

"A new headline. All-Pro Raider Goes Prim."

He grabbed her and kissed her hard on the lips, then just held her close. The sweet fragrance of her loosened his knees, and she threw her long arms around him to hold him up. "They're really getting to you, aren't they, Brad?"

"What are you talking about?" he said sharply.

"Okay, if you don't want me sticking my long nose into your affairs, I won't. But I'm available. Call whenever you're lonesome or whatever." She turned to leave.

He regretted his angry reply to her, started to articulate his tongue in order to apologize, thought better of it. He had a date with Golo and he didn't want to be late. The light said to walk, so he walked, alongside him a young black woman and her seven-year-old child. The mother looked care-worn, harried. Clutching her daughter's hand, she hurried the child along, seemingly oblivious to all about her. He wanted to tell her to take it easy, it was slippery. The child, too, looked unhappy, not properly dressed for such wintry weather, with her free hand holding her light cloth coat tight to her thin body. The red light held, cars braked cautiously. One car did not, applying its brakes too abruptly. It lurched, then hurtled foward, fishtailing through the red light.

Bradovich, still possessed of the superior reflexes and acute peripheral vision shared by all great athletes, sprang hurriedly to life and, as the careening car leaped straight at him, sidestepped. It missed him—just. Completely out of control, the onrushing four-wheeled missile sideswiped the black woman and child who were slightly to the right and rear of Bradovich. All traffic stopped, but not the rogue car. It came under the control of its driver and went roaring down the icy street. As Bradovich turned toward the fallen woman and child, he made a quick mental note of its license plate number.

The stricken woman lay on her back, the child beside her. Bradovich sat down on the ice and, afraid to move the woman, her back might be broken, he took her hand in his and felt her pulse. Feathery light, but alive. The child was unconscious but her gasping breath informed him that she was still alive as well. A crowd had gathered, and Beatrice Holden who'd seen the whole thing ran forward to help Wizard. As she attended the little girl, who suddenly sat up and stared vacantly, Bradovich wiped the mother's brow with a handkerchief. Her hand was icy cold to his touch, her brow as well. He yelled out, "Someone call a goddamned ambulance!" and someone yelled back, "We already did."

Bradovich continued to wipe the woman's cold brow, afraid to move her, wanting to pick her up. She opened her eyes wide in terror, sightless, and suddenly grabbed his hand with terrifying strength. Then went limp, and Bradovich knew she was dead. In death, her hand still held his, and he permitted his fingers to learn about death from her, and it was not much different from what he had learned from Valerian. Cold, bony, in death as in life, poor woman. He sat like that until the ambulance attendants came to take her away, having to pry her fingers from his hand. "Goodbye, dear lady," he said, as they gathered her up and carried her away.

Beatrice held the child close to herself, humming a lullabye of sorts. The little girl had come awake, but just stared mutely, still in shock. She was a sweet-faced child, with a delicate broad nose, lotus lips, and eyes which were large and blank now. As he stood behind and close to Beatrice, one hand on her shoulder, the child stared at him, and seeing someone she thought she recognized, she raised her arms to him, as though asking him to take her. He opened his coat wide and Beatrice transferred the child to him, herself stepping close so that the little girl received warmth from both of them.

Staring up at Wizard, the child raised her hand and with her fingers traced his eyes.

"I'll catch your tears," she said, speaking for the first time.

He had no tears, but he said, "Yes, child, catch my tears."

The body heat from Beatrice and Wizard thawed the little girl's shock and like a baby suddenly come awake she screamed, "Mama! Mama!"

Her mama was now on her way to the city morgue where she would be refrigerated until some relative could be found to identify her. Shortly, a police matron arrived and took the wailing child from them. Bradovich gave an attending officer the license plate number of the hit-and-run car, then continued, Beatrice alongside him, her hand on his elbow, to the bus stop.

"I'll see you later," he said, kissing her on the lips as the bus approached. They were now joined. She smiled sadly and waved to him, as he saluted her from inside the bus.

After he sat down, he looked around the inside of the

vehicle. It was only half-filled. Ordinary. Plain. The same as always. The woman had died. The little girl had become motherless. Nothing had changed. Still in the sadness there was a new warmth. Beatrice. He couldn't but smile to himself.

As Bradovich entered the cavernous lobby of the midtown glass tower, he saw a familiar figure reflected in the glass door. Tall, square-shouldered, blankfaced. He smiled grimly. They were interchangeable, the bastards. Could that car have been meant for him? No, no, he was getting paranoid.

*B*radovich was two hours late for his appointment with the Gimp. He hoped that his friend had been able to wait. Golo was on the move endlessly, almost as if he were afraid to stand still. Afraid his shadow would catch him.

Sam Rabinowitz, who was the president of Intercontinental Enterprises, must have been obsessed with the color pink, for the walls, the ceilings, the wall-to-wall carpeting were all in that hue. Even the beautiful young men and women who served as receptionists and secretaries were pink-cheeked. By the color of their skin, Bradovich was willing to bet all the young women had pink areolas, pink cunts, and pink assholes, and the young men pink cocks, nipples, and assholes as well. Ah, there was one to make a man think. He was ogling a young Indian woman, gold star on forehead, diamond-studded nostril. Nothing pink about her. Gorgeous head.

"My name's Bradovich, I'm a sculptor. I'd love to do your head."

"And I'm Medusa. Get on line."

"I'll try again."

"Do that."

They smiled at each other, and as she went ahead of him he inscribed every inch of her on his mind. "The female of the species is superior," he hummed.

Rabinowitz, the son of immigrants from Bucharest, had risen from poverty to great power in the entertainment world. His list included the most successful actors, directors, producers, and writers in the world. Of them, Golo the Gimp was among the most sought after, and Intercontinental Missile, as some called it, kept an office just for his personal use. In addition to his work as a performer, Golo was a board member of many charitable institutions, and during election campaigns, whether local, state, or national, was called on by politicians whom he endorsed to make speeches on radio and television. Prescient in political matters, as in most else, the Gimp had never backed a loser.

When Bradovich asked the pink-cheeked receptionist to tell Golo that he had at last arrived, the young woman wet her pink lips with a pink-tipped tongue, then said, "I'm awfully sorry, sir, but Golo"—the Gimp used no other name, and no one seemed to know if he had one—"couldn't wait any longer. He had to catch the Concorde for Paris, then a jet to Oslo where he has to perform tonight. He'll be back in three days and will call you. Here's a note he left for you."

Dejected, Bradovich sat in one of several Barcelona chairs—not pink—bequeathed by Mies van der Rohe to Golo who had once made the great architect laugh and cry simultaneously with his most celebrated act, a pantomime in which Golo played both Hitler and Stalin having tea and crumpets together on a pile of bones.

82

"My dear Wizard," the note began, "I'm unhappy to have to tell you that your life is in danger. I have already spoken on your behalf. My contact said he would think about it. He owes me one. I once spoke on his behalf to an even higher authority. There always is one, isn't there? You have a good chance, though, the odds are in your favor. I'll keep trying. You're my friend, and Floriana's as well. I love her and will win her, you'll see. You've helped—she told me how you forwarded my suit. I want a child and I want one with Floriana. Her and her goddamned brother. She does love me, however, and for myself. As you said to me, she also loves her independence. Do love and independence conflict? She told me despite the fact that I'm a short gimp I have the biggest *pinga* she has ever enjoyed." A man and his cock, a king and his sceptre. Silly Golo, as if his friends fared how little or big his cock was. Little man, how big you are. The better to beat you with, my dear. "Be well," the note continued, "and try not to worry. Hang tough, old buddie. The Authority can't but respect that. Make a strong effort to control your temper, that only exacerbates the problem. By the way, I ran into Beatrice the other day. Don't hold back, you idiot, she has great affection for you, and is worried about you. It is over three years now since poor Valerian died. It's time, isn'tit, you stubborn mule, that you found life with another? See ya, G."

Bradovich relaxed in the Barcelona chair, of two minds, perhaps three. How many minds are there? First Golo tells him his life is in danger, then he tells him he has a good chance—which is it? On the other hand, Golo had already intervened on his behalf, still everywhere he went he was shadowed. No violence, they say, then beat the shit out of him. Naked he lay before them, the swine. The welt on his belly still hurt. Perhaps Golo's contact wasn't high enough in the hierarchy of The Authority. Time would tell. And

Beatrice was worried about him. Well, he already knew that. He remembered how he'd felt the strong beat of her heart as they held the child between them. They were already joined. She was a nifty lady, no bullshit with her, and that ass—how he'd love to eat that peach. Besides, he couldn't go to Floriana any more. Her, Golo, it was getting to be too damned queer. He had absolutely no interest in Golo's *pinga*. Bradovich laughed aloud. "How many minds is that?"

On his way home, he stopped in the precinct house in his neighborhood. No, they hadn't apprehended the hit-and-run driver. The fact was that the license plate number he had reported to the police was non-existent.

"Impossible," he said. "I caught that number and made sure I memorized it, and I happen to have a very good memory. May I see the number you checked out?"

It was shown to him by the policewoman in charge. She was a tall, willowy brunette with saucer blue eyes and a large nose.

"Yup, that's it. I'm certain!"

"I'm sorry, pal," she said very politely, her saucer eyes wide, her big nose quivering over her thin lips, "no such number appeared on the computer."

"Could it have been a forged plate?" Inwardly, Bradovich trembled at the thought. Could it be them?

"It's happened," the officer said, rolling her large eyes. "These days anything can happen, and does. What people do, they do."

Startled, Bradovich looked sharply at her. The trial in the park—he'd tried to forget it as a hallucination.

She shrugged her narrow shoulders, smiled grimly under her overshadowing nose. "If we find that bastard—it was a man driving, wasn't it?"

84

"It certainly was."

"We'll hogtie him for sure and really do it to him."

"Do what?"

She shrugged, her saucers rolling upward as if to say, What a jerk.

"The poor woman, and the child left an orphan."

"She wasn't left all alone, we found the father."

"I'm glad to hear that."

"Yeah. Well, thanks for your help, pal."

*T*hat night, Bradovich kept waking from a deep almost drugged sleep. He was certain now that the driver of the runamok car was aiming at him, his intent execution, and because he'd been alert enough to sidestep its murderous approach that poor black woman had been killed, the lovely child made motherless.

Golo's note had said, ". . . your life is in danger." Why? The Authority's operatives had said he was only under surveillance. Now this attempt on his life. Its power was absolute and absolute power is arbitrary power. It can do any goddamned thing it wants when it wants to do it. It super-cedes law, order, rules, morality, commandments. Even its own!

Bradovich rose from his bed and padded into the kitchen where he lifted a bottle from the cabinet. He threw back a two-ounce shot glass of corn mash neat. Then an-other. It burned down his gullet, and then spread its warmth throughout his belly. His father had known of this particular cure for fear. Had he been harassed by The Authority too? To make certain the cure worked, Bradovich knocked back another shot.

Returned to bed, he slept a black sleep, dreamless.

When Wizard awoke in the morning, he felt as if he'd spent the night breaking rocks with a sledge hammer. In modern prisons did they still pound rocks in penal servitude? Or would he be sent to some Devil's Island where he'd be brutalized by the tropical sun and the *lumpen* guards? Would their faces be faceless, polished like the heads he sculptured or would those of the victims be blank, burnished by the sun and blanked by time? Why was it when you thought of prison—or hell—you only envisioned what you commonly called horrors such as whips, chains, the rack, fire, brimstone? Banalities could make for a worse hell than those.

Mrs. Kastner was at her window again. He could always count on her to bring him back to reality. She was singing in that cracked voice of hers.

I'll never haunt you.
I'll never lie. . . .
I'm a one-man woman in a two-timing town. . . .

Her husband was the biggest skirtchaser in town. He was so bad, Floriana wouldn't allow him in her door. "I never trust a vagabond prick, you never know what it'll bring you." Valerian had once told him, her husband, "I don't want you to hit him, just nail him to the wall with a couple of spikes through his groin." Beatrice Holden had had to call Slim, the doorman, to get Mr. Kastner away from her door. For Mrs. Kastner, wasn't that a hell worse than chain gangs? Frankie, his first wife, had been just a little better than Mr. Kastner, had made the rounds of most of the men in their circle. Before she'd gone off with the glitzy painter, she'd had a child with Bradovich. A sweet little boy born with an esoteric

87

blood disease. The specialists had done their best, but he died anyway, his fragile veins hemorrhaging. The crib was soaked with blood. The child was buried under a simple stone—his name in simple lettering; his years two. He'd had red hair and bright green eyes.

After their son was buried, Francine seemed to change. Became quiet, what one would call a normal person. She sang sad folk songs. But not for long. Class will tell. Soon she was driving Bradovich crazy again. A two-timing woman in a two-timing town. I was nuts about my father, she said, as if that excused her behavior. Her father was long dead. It's the life force in me, she said. Another excuse. I'm a bohemian, not a bourgeoise, she said. More shit. Phony bitch.

She'd left him just in time, or he would have killed her. He'd even gone to the library to read chemistry books in search of the perfect poison. Wanted to throw her off the roof after luring her there to catch a breath of air in the dog days of summer. Once he almost drove the rented car into a huge tree, glad to go with her to escape their trite hell. Another time he sharpened the meat cleaver with the thought he'd submit her head to the axe like Anne Boleyn's. Talk about the banality of evil. The tabloids were full of it. How silly it had all been. Why hadn't he just walked out on her, kicked her out, whatever? Because he'd been a silly fool. He'd felt sorry for her—she was a frightened bunny caught in the glare of life. My God, how she'd looked for escape. Now when he ran into her he laughed like hell. She'd become a damned fine painter. No accounting for that. That also was arbitrary. She was a disorderly person. On canvas she created an order of ineffable beauty.

The Fucking Authority, who the hell does it think it is? Maybe he ought to buy a gun. If he shot one of those pokerfaced cocksuckers, the public scandal that followed might reveal who The Authority was. Was it a public organ or a vigilante group who chose its victims at random just to show its power? What exactly had he done to deserve this? Which particular sin, or was it a combination of sins? Relax, bud, just relax. Wait to hear further from the Gimp.

Bradovich turned and rolled under the blanket. He straightened out, curled up, flung his blanket off, pulled it back over himself. He slept in a deep hole. A pit. A grave.

The sun in his eyes and the need to urinate woke him. When he jumped from his bed, he heard a whistle followed by "Wow!" It was Mrs. Kastner, now grinning. He covered his piss erection with his hands and ran to the toilet. Jesus, what a looney broad.

It had snowed again during the night. A blizzard. The city was brought to a halt. Absolutely no movement. The snow plows and sanding trucks worked round the clock. It snowed for three days. Where, he wondered, were his shadows? Were they outside his door? He looked, and they were not there. Were they downstairs in the lobby? He phoned Slim. "No, nobody's here ain't supposed to be here. You're nuts, Brad, just plain nuts."

Perhaps they were out in the street freezing. I hope their balls get so hard they can be used for bowling.

Bradovich remained in his apartment, holed up, until his food ran out. I've got enough fat on me to last a good month. Beatrice phoned. "I'm sorry," he said. "I'm working on something, real obsessed," he lied, "can't think of anything else." "I understand," she said. He fasted for several days, only drinking water. Began to feel otherworldly, light as air, but the headache was so bad tears flowed from his eyes.

Just as he was getting ready for the looney wagon, Golo phoned. "Things are looking up," he said, his voice a growl. "There's a good chance, kid." A good chance for what he didn't say. "Floriana says you haven't visited lately."

89

"Floriana's your girl. I'm beginning to feel like a rat."

"Don't tell her that, she'll be furious. She'll say, 'You see, Gimp, you're interfering with my life.'"

"With this weather I daresay she hasn't been very busy."

"She's too damned busy so far as I'm concerned." He sounded sad to Bradovich.

Wizard felt sorry for him. "Like you, she's an entertainer. On stage, she thinks only of her audience—offstage, she thinks only of you. I'm sure you understand that."

The Gimp laughed. "Thanks. It seems, my dear Wizard, you become wiser and wiser with time."

"Yeah, wiser."

"By the way, it might cost a little, sometimes a contract has to be made in these matters."

"Like hell! No contract! Not a dime! Whatever I've done, I've done. I won't bribe anyone to get off. Never have. Never will."

"Okay, okay. I understand you have some pride in this matter. Leave it to me."

"Will I be seeing you one of these days?"

"First chance I get. Right now I have a command performance at the White House to prepare for—have to collect a whole troupe, an old-fashioned vaudeville bill for the incumbent tap dancer, so it will take a little time. Keep punching, old man."

"You know I appreciate—"

"Cut it out, we're old friends."

Bradovich stood in the middle of his bedroom floor and did calisthenics. The headache had disappeared. Could eat a horse. Two horses. He laughed.

His back was toward the window, but he could hear Mrs. Kastner opening hers.

"That's a great ass you got there, Wizard."

He ignored her.

"I bet that old Bradovich cock can still get hard as a rock."

He turned around to show her his smile and stood on his toes to permit her to catch a glimpse of his limpness.

Mrs. Kastner stuck her thumbs in her ears and wagged her fingers, her tongue lolling from her open mouth.

*T*he sun was burnt orange.

A mid-winter thaw. The gray was overwhelmed. People in the streets smiled at one another, shading their eyes from the sun. The ice on the boulevard's lanes melted, puddles formed, people laughingly jumped aside as car wheels splashed.

Bradovich's windows glittered gold. His rooms became barred with motes dancing in burnt orange light. He could not reject the sun. He showered, shaved, splashed fragrant scent on his cheeks, dressed in crispy white boxer shorts and T-shirt, a pair of forest green corduroys, a subtle plaid shirt, a sturdy pair of walking shoes. Hatless, a padded but light-as-a-feather jacket, designer labeled, zipped a quarter way up, he triple-locked his door and descended via squeaky lift to the lobby. His mailbox was packed tight with junk mail from Aardvark to Zeta Zinc all of which he threw into the trash basket.

Hand up to shield his eyes from the burnished sun, he

strode a few steps next door to the coffee shop which deserved three Michelin toques for its breakfast—the coffee strong and hot, the toast done a uniform tan, the once-over-lightly eggs exactly as he liked, and the homefries, ah, the homefries! The short order cook was polite and minded his own business. A perfect breakfast combination. There was nothing in the world Bradovich detested more than forced geniality when he ate out, especially at breakfast. He ordered two portions of everything.

He wiped the last of the egg from the plate with the last bite of toast, chewed it slowly and lovingly, emptied the cup of its coffee, turned down an offered additional cup, paid the bill, left a big tip in the glass near the cash register, and strolled out feeling as if he owned the world.

Saw one of his shadows. Gave him the finger and a loud "Screw you, bub!" Those who heard him, joined in his loud laughter. Stoneface smiled smugly. Strange. Something new.

Golo and the sun had replenished his energy. He entered the lazy flow of pedestrians under the sun-drenched uptown side of the street. As he passed Dominic's Italian Delight, he saw to his astonishment that the Cosmic Travel Agency was open, and without missing a heartbeat—forgetting momentarily both his shadow and The Authority—he stepped out of the pedestrian stream and into the sun-splashed storefront.

Sitting before a young handsome woman in shirt and tie, he explained his purpose. She punched buttons on a computer, then dialed a number. "AOK."

She smiled for the first time—had the whitest, straightest orthodontist-manipulated teeth he had ever seen. Looked like an ad for dentures. He returned the smile—two nonremovable bridges, four capped molars, and one removable splint. The flight tickets made out, the check paid, the packet placed in the inside pocket of his jacket, Bradovich left the agency.

The sun's rays had crossed the street, so he went in that direction. The stream downtown moved at a fast clip, life must go on, there were things to be done. Keeping in step with those about him, ignoring two female stonefaces, his eyes squinting against the orange sun, Bradovich walked up the back of someone in front of him. The man turned to say something nasty, saw Bradovich's hulk, thought better of it. "Sorry," Wizard said, caught in a logjam before a building being gutted before restoration. The sidewalk was roped off, and pedestrians were detouring into the street around it. Someone screamed, "Watch out!" and people scrambled away. As Bradovich shuffled his feet, something or someone struck him behind his knees and he went tumbling over the rope and down on his hands. As he hit the ground, he turned his head to look upward and saw a huge piece of concrete cornice falling toward him. And beyond it the grinning face of the brutal twin. Wizard, still possessed of his senses and reflexes, rolled quickly aside as the huge chunk of concrete pounded the pavement where he had been a mere split second before. Fortunately, he had covered his face and head with his arms which along with the rest of his body were pelted with concrete fragments. The street and sidewalk were crowded with pedestrians and shoppers, their mouths agape, there larynxes mute from sudden fear. Bradovich sat quietly trying to control his breathing, his eyes closed, his mouth slightly open. A near miss. That was close. Yes. Yes, oh, yes, it, they, The Authority meant business. Lucky this time yet again.

The street's silence abruptly came to an end, and the boulevard resumed its business. A couple of workmen from the building stood near Bradovich and asked him if he was all right. "Yes," he said, standing at last.

"For an old man," one of them said, "you sure moved damned fast."

"Forgot my age in the excitement."

"I don't know how you can smile," the workman said.

"Am I smiling?"

"You sure are, mister."

"That must be the way I show hysteria. I'm scared shitless."

"Can you manage okay?"

"Yes, thanks, bud, I think I'll be all right."

The men helped him clean off the debris and powdered concrete, and then, saying thanks, Bradovich continued his walk, hardly caring in what direction he went. Shortly, as he was passing the Epicurean Coffee Shop, he decided he might as well have a cup of Viennese coffee *mit schlag*, it would return some of the energy that had just been drained from him.

Inside the shop, he removed his warm jacket and hung it over the wire back of the old ice cream parlor chair. He had to balance himself carefully on the chair since he overwhelmed it. They make these bloody things for skinny ladies. The coffee with whipped cream as well as the strudel he had also ordered satisfied his need for sweets, even reassured him, and he sighed with relief. Yes, he was still alive. He sat for a few minutes longer, thinking how nice it would be to have a small Dutch cigar, but he'd given up smoking five years before, when Valerian, still alive, had begged him to. Leaning back carefully—he worried about the fragility of the chair—he relaxed enough to enjoy the peace and the pleasant tastes which lingered in his mouth. Fear sneered at him, but he pretended to ignore it. He was still alive—by a hair. Those dirty hypocritical bastards. He had to be constantly on the alert. He sat there, in no hurry to move, his legs still weak, and began to ogle the long shapely thighs of a woman who was enjoying an eclair at a table nearby. Her short short skirt was hiked up to her panty-hose covered crotch. Made him think of Beatrice with her cloud of red hair. She was tall, too, with long shapely legs and neat ankles. She was being generous with him, why was he being so miserly in return? What was he waiting for? Call her, make a date. The tables in the small shop were close to one another, and several times he was jostled by customers going to and coming from tables. He was more or less oblivious of them as he continued to ogle the young woman's sensual legs and mysterious crotch and mused about Beatrice's wide and what could only be rewarding smile. A familiar tall figure just slipped out the door. Frightened, he reached his hand into the inner pocket of his jacket hanging begind him on the chair. The plane tickets were gone. Frantically, he searched around the chair and under the table. That goddamn black-suited figure. Damn them! He threw several dollars on the table and rushed out of the shop. Saw the high square shoulders down the block, and headed for them on the run on still shaky legs. The streets were crowded with promenaders under the sun, and his vision was blocked. By the time he hit the corner, the operative was gone, had disappeared into the sun.

Rushing to the Cosmic Travel Agency to cancel the tickets and obtain new ones, Bradovich found it closed. Will reopen tomorrow.

Bradovich ran home, scattering people all the way. Neither the sun nor the smiling walkers helped now. In his apartment, behind locked door with window shades down against the day, he sat hunched in a wing chair. Golo was gone, it was senseless to phone him. There was nothing he could do but wait. It would be useless to go to the district attorney, to a lawyer. No one could help him except the

Gimp who understood better than anyone how the world worked, its power sources. He was a fool. Why not follow Golo's advice? If money was needed, if palms had to be crossed, he would cross them. Who the hell was he to be so incorruptible? Power always had to be fed, favors bought. With gold, with human sacrifices. A quack god, a fucking charlatan, power bestrode the earth, grinning greedily, demanding your obeisance and a greenback.

Bradovich sat stiffly in his easy chair. A hive broke out on his face and he scratched it. Then one irrupted on his thigh. He scratched it. He began to feel feverish, sweat oozed from his pores. Shame washed over him with its acid. With all the talk of child abuse, now a grown woman in late life she'd gone to the authorities. Bradovich, the well-known sculptor, had plunged his hard eighteen year old hand down her middie-blouse to cup her nubile breast when she was but a thirteen-year-old schoolgirl. She'd shrunk back from deep embarrassment, cried from fright, traumatized for life. He'd run out of her house, leaving her brother in the next room, his friend, to wonder, to question. Another hive sprung to life on Bradovich's chin. He scratched it ferociously. He would be exposed, perhaps was already exposed, a figure of ridicule like a flasher. He'd done it a second time, to a babysitter, many years later, couldn't keep his hand off her budding girlish breast, walking the child home after returning with Val from an artists' ball half drunk. My God, the shame! The girl's father, another friend, had confronted him and he'd replied with a guilty grunt, his voice box paralyzed. Now his errant passion had come home to haunt him. His lips swelled from fever blisters. His back abruptly irrupted with a thick carpet of hives. The sweat burned. He scratched wildly, writhed against the back of the easy chair. His armpits exploded into hives. His crotch. He lay himself on the floor and violently rubbed his body against the nap of the rug, acid sweat oozing. He was being burned alive, the shame, the itching. He yelled curses at himself. At Valerian for daring to die. At his mother. His father. Francine. Cursed God, if there was one. But mostly he cursed himself. The shame, the shame. Now he understood the harassment, the threats to his life. The Authority laughed uproariously.

The phone rang. Rang and rang. He refused to answer it. It continued to ring as if the ringer knew he was at home. The hell with it. It must be them. Still the phone rang. The kids, his kids. . . .

It was Beatrice. "I knew you were home, Bradovich. I heard what happened this morning. I'm ready to give up on you. I was sure you'd call me, I've given you enough hints to—to—"

"I know. I'm sorry. I'm—"

"You're in trouble. So what. Who isn't in trouble of one sort or another?"

"Right now I'm dying from a bad case of hives."

"Hold still, I'll be right down. I know just the cure for hives. I've had them myself."

Despite himself, Bradovich smiled. Beatrice was exactly what he needed. When he opened to her ring, she pushed past him and headed for the bathroom, saying, "Get me your box of baking soda, it's in the refrigerator."

"How do you—"

"I'm omniscient. You think I'm a fool, but I'm not. I'm pretty damned smart, Bradovich. They don't pay me thirty bucks an hour to edit books by Nobel laureates for nothing."

He didn't argue with her. When he brought her the box of baking soda, the tub was already half filled with steaming water. "Get undressed," she ordered, emptying the box into the tub.

He hesitated.

"Don't be an idiot."

He undressed.

"Now get in the tub."

"It's too bloody hot."

"It won't kill you, Bradovich."

He lowered himself into the scalding water, biting his lips to suppress a howl of pain.

She lowered the shade on the small window. Turned out the light, went to his bedroom and returned with the portable radio. Plugging it in, she dialed the classical music station. It was in the middle of a Beethoven piano sonata. "Close your eyes, relax," she ordered.

Bradovich felt like a child, and enjoyed every second of it.

Beatrice closed the door behind herself silently, and in a few minutes left the apartment.

He fell asleep in the tub. When he awoke, the hives were gone.

Later, on the kitchen table, he found a note from Beatrice.

"Look, Bradovich, I'm not a cunt in search of a cock. I'm a woman looking for a man. Still old-fashioned that way. You're not a dope, I'm sure you understand. Valerian must have taught you something. I'm lonesome, aren't you? Why are you so slow? Make a move, Wizard, I've done the best I could. Our love was made in heaven, ha ha. Beatrice."

He wanted to raise the phone then and there and call her. To thank her. To make a date with her. But he didn't. With Floriana it was all very simple, no strings attached, an easy transaction. With Beatrice, Valerian interfered. Why should it? Valerian, dying, had told him not to wait too long, to get himself a woman quickly before stasis set in. "Once you get into a one-way rut, Mister Bradovich, you will never find a way out of it." Yes, she'd known him very well.

He could see her in sharp focus. It wasn't just that she'd been a goodlooking woman, it was the way she'd placed herself in space.

"Can't you move a little closer, Val?"

"We're close enough."

"I don't think so."

"If we get any closer, there won't be anything left for me. You're too big, too strong, too stubborn, and I'm too compliant. I need to protect myself, and I do that by giving you the space you need, because I need the space that's left for me. You see, I'm selfish."

Who was it who'd been selfish, she or he? When he'd wanted her close, he'd move into her space without asking, and she'd move away. "What's the matter, Mister Bradovich, are you a baby, need mama to feed on? You move in and out as you please. I want none of it. I demand my rights."

She'd been right, of course. He had taken his rights for granted, hardly giving her hers.

But when life became really tough, she was there in her hands-off sort of way. He wished she were here now that he was in trouble with The Authority. She would have known how to handle it, have cooled him down. He wasn't sure Beatrice could be that calm; God knows, she was asking for it. And he recognized that Val was Val and Beatrice was Beatrice, just as Bradovich was Bradovich and not John Doe. He liked Beatrice, wanted her, yes, even needed her. She was willing, she thought she knew what she was doing. She wasn't a kid. So why not? Because The Authority said no. How do you know that, Bradovich? They said you were on your own recognizance so long as you remained within the city limits. Fuck 'em! Oh, yeah, try it.

*C*osmic Travel Agency had gone out of business. Its storefront windows were papered over. A cab dropped Bradovich off in front of the offices of the airline which had booked him. The clerk, a handsome blond man with broad square shoulders who stared at Bradovich with unbelieving eyes—he looked like a white leopard— pushed buttons on a computer. Words flashed on the screen.

"Are you certain you booked passage on our line?" the white leopard asked, a sneer scribing the corners of his sullen feline mouth.

"Yes. Through Cosmic Travel."

"There's no record of it. And Cosmic's closed down."

"I don't give a shit. I paid for those tickets," Wizard shouted.

"There's nothing we can do about it," snarled the leopard. "You're in trouble, mister."

"So are you, you prick." Bradovich repositioned his feet slightly, shifted his eyes to the right, poleaxed the blond cat with his left. As other employees of the airline came running, an armed guard among them, Bradovich walked out, hailed a cab, and returned home.

As he entered his apartment, the phone rang.

Golo spoke to him quietly, calmed him down. "You can't go around doing things like that, you're just making it worse for yourself."

"Let them get their investigation over with. Life or death, who gives a damn, but I'm getting awful tired of it."

"Hey, relax. I have an appointment with a big shot on Sunday, two days, the Oak Room; if you'll permit me, I'll try to make a contract."

"Why do you insist on trying to corrupt me, Golo? If you kneel to power, it takes it as an invitation to lop your head off."

"I'm just thinking of your welfare."

"I'm aware of that. And appreciate it. But I'm not going to change at this stage in my life. It just isn't worth it."

"You're a stubborn ox."

"You're right. That's my forte. Stubbornness. It's the reason I'm still alive."

"I have to agree with you. That's why I admire you. In the meantime, avoid unnecessary quarrels no matter how frustrated you become. Forget the coast to visit your children for the time being. In fact, they should be coming here to visit you now. Have you told them?"

"No, of course not, I don't want to bother them with my problems."

"I don't see why not, they've bothered you with theirs. All right, you don't leave me much room, but I'll do the best I can."

"I know you will."

"*Hasta la vista.*"

"Sure."

For the next few days, Bradovich avoided all contact with those who might cause him distress. When he went shopping at the market, he made certain to be polite to everyone, other shoppers, clerks, even the checkout girl, a contemptuous young woman who felt superior to her job and her customers. How could you possibly be anyone if you shop in this dump was written all over her face.

Once on the street, he was very careful. Should he accidentally be bumped or himself bump someone, he was, if anything, excessively apologetic. Almost bowing. I'm becoming a character out of Gogol. Once he changed his mind about his destination and spun quickly on his heel and stepped smack into one of his shadows. "I'm sorry," he said, "I don't mean to interfere with your duty." The stoneface was so taken aback, it spoke. "That's perfectly all right, forget it."

Perhaps the Gimp had prescribed the right medicine. There were just him and his shadows, otherwise peace.

Early one evening, after a good six mile run in the park—a duo of shadows ran, too, easily keeping pace behind him—as Bradovich strolled on the boulevard toward home, a cab screechingly pulled over to the curb, and the cabbie shouted to him to stop. He did. In the rear of the cab a woman was lying on her back, screaming. She was, it turned out, in the extremity of labor. The cabbie, a small man, was crying, almost hysterical. Bradovich spoke quietly to him, told him to go into one of the nearby boutiques to ask for clean towels as he opened the back door of the taxi and leaned in as far as he could reach.

The woman, her skirts thrown back over her inflated belly, her knees spread wide, was howling curses at the top of her voice.

"Keep screaming," he said to her. "Scream your bloody head off."

She hardly needed his advice. The air was blue around her.

The cabbie was back shortly, pulling at Bradovich's pants. Taking the white towels from him, Wizard again reached into the taxi and lifted the woman's behind and spread the towels under her. The baby's head, he saw, was ready to break through. He had retained one of the towels and held it ready. "C'mon, one really big one," he shouted at the woman.

She let go an explosion of curses, at him, at the world, let go a blast that shuddered the cab, and soon the pressure on her lower abdomen sprung the baby loose. It shot out with a spurt and was caught deftly by Wizard.

There were hurrahs, wows and olés from the large audience which now surrounded the taxi. Fortunately, at that moment a police cruiser skidded to a halt and out poured, literally, two mammoth cops. They had the necessary instruments kept in readiness for events such as this, and one of them expertly cut the umbilical as Bradovich held the bawling infant in his arms. A boy it was, who, he was sure, would bawl and laugh his way through life, The Authority be damned.

He left the grateful woman and her child to the cops and hurried toward home, his jogging suit filthy with blood and afterbirth. In the midst of all the shit, there was always this little cliché of a miracle, a baby was born.

Outside his apartment building, he encountered Beatrice under her red cloud. He would have to start wooing her just so he could induce her to change the way she did her hair. When she saw the blood on his jacket, she visibly suffered. "Did they do that to you?" she asked in a choked voice.

He wanted to ask her what "they," but decided to let her think as she wished, and said nothing. She obviously

104

thought he was too shaken to respond. Grabbing him by the elbow, she led him toward the elevator. "I'll clean you up," she said in a husky voice. It created a stir in his pants.

Once in his apartment, Wizard told her about the birth of the baby in the cab, and she sighed with relief. They were in the dark hall, and stood suddenly mute, she biting her lower lip, he wetting his lips with his tongue. Their eyes held, did not run to hide. Wizard took her by the shoulders and she swung her long slender arms around his waist. Her breath came fast, his sounded like a locomotive, and they were on the carpeted floor, Beatrice's knees high over her head. It was over much too quickly, and Bradovich apologized, "Just like a kid on his first time out."

"It's all right," she whispered in his ear. "It was great, besides you don't have to hit the jackpot every time."

"You're a peach," he said in her ear. "A real peach."

He asked her to go to dinner with him at the Hunan Inn, but she said she had a deadline she had to meet. "Tomorrow, then," he asked.

"Sure, of course."

At the door, they embraced and kissed, a long, soft and warm kiss. His hand slid under her skirt and he tenderly pinched her lubricious cleft. It made them both happy.

He undressed quickly, throwing his bloodied clothes in a heap on the bathroom floor. Then he took a shower under very hot water. As was happening more frequently lately, midway it changed to tepid, then to cold, then to hot again until he became tired of fiddling with the faucets and pounded the wall with his fists, which of course altered nothing, except that Mrs. Kastner pounded back from her side. When he pulled the shower curtain aside and stepped out of the tub, Mr. Brutal Twin was standing there with an ugly grin on his stony face. Before Bradovich could draw back, The Authority's operative jabbed him viciously in the solar plexus. The pain was so sharp Wizard sat quickly on the tile floor, holding his stomach and gasping for breath, tears burning his closed eyes. He thought he was going to pass out, so he lay on the floor and pressed his head to the cold tiles, the smell of childbirth from his jogging clothes in his nostrils.

Slowly, the pain receded, and he began to breathe normally again. He merely lay there, alternately frightened and furious. They'd made him pay for Beatrice, he was certain. Call Golo, tell him enough, to make a contract. Bradovich sighed deeply. You live a lifetime fighting corruption, but you can't keep ahead of it, it's always there, getting at you little by little. In the end, it beats you down, wins. The fucking victor!

Later, after he'd dressed, he phoned the building maintenance department and informed the man who answered about the hot water problem.

"Nobody else has complained, Mr. B. It can't be just your shower, for God's sake. Maybe your body temperature's changed or something."

Suppressing the desire to yell, Bradovich said, "You're right, it must be something else, not the plumbing. Yeah, maybe it's me."

"That's right. I'm glad to see you're being reasonable for a change."

"Yes, it's time for a change. Sorry to have bothered you."

"Okay, okay, I got other things to do."

To prevent himself from slamming the phone down, Bradovich counted to ten, then hung up slowly. As he turned away he could smell Beatrice in the house. A strong musty odor. She wasn't a sugarbun. Thank God.

Several days later, early morning, Bradovich was going from his bathroom to the kitchen and heard a rapping at his door. It was rare that anyone would arrive at his apartment without first having been announced from the lobby. The rap was insistent, though dainty.

He undid the chain, unbolted the lock, swung the door open. Before he could say a word, the old woman standing on his threshold brought a thin finger to her pursed mouth to indicate silence. She was a little lady, white haired, dressed in a blue house dress over which she wore an old patched tan cardigan, and on top of that a gray overcoat worn capelike, empty sleeves dangling to below her waist. He furrowed his brow, now who the hell is this, then opened his mouth to speak, but she whispered "Sshh," and against his will he closed it.

Like a practiced mime, using her hands and body, the old woman asked whether she could enter. One of her hands,

Bradovich noticed, clutched a small wood-framed slate. Shrugging his huge shoulders, he stepped back and made room for her in the hall. She took the doorknob in her bony fist and closed the door so quietly the click couldn't be heard, then fastened the chain. She made no movement to indicate she wanted to go any further into the apartment, so he stood near her, a giant to her slightness, wondering what the hell it was she wanted.

The hallway was dimly lit, and it seemed she wanted a better look at him, because she stood on her toes and raised her free hand to the nape of his thick neck and gently drew his head down to her eye level. Do I have to go along with this silly charade? Yes, sure, go ahead. Let's see what happens.

Her pearl gray eyes stared into his ice blues, stared almost as if she wanted to hypnotize him. Tears formed at the ends of her eyes and slowly slipped down her waxlike cheeks.

Bradovich resented her tears and loosened her hand, not gently, from his neck and stood erect, staring down at her sternly. Again he opened his mouth to speak, but the woman shook her head violently and brought her thin fingers swiftly to her pursed lips. It was an order, and Bradovich responded to it like a good soldier, and wondered why.

The woman withdrew a stick of chalk from an inside pocket and wrote a few words on the clean slate. Her handwriting was so hurried and nervous, he found it impossible to read. He took the slate from her thin hand and wrote, "Come into the kitchen, I'll give you a cup of tea, that'll calm you down until you can write more legibly." You have to be kind to the looney.

She smiled weakly and nodded her head. He took the coat from her narrow shoulders and hung it in the hall closet. Once in the dinette with its very bright light, he saw that despite her age she was a beautiful woman, her cameolike

108

features finely etched, and her eyes, in the dim hallway seen as pearl gray, were gray green, so clear that they reminded him of an Adirondack lake during a summer rain when Val and he had vacationed there with the kids. Though impatient with this charade, he found himself deeply affected by the sad look in her eyes. He would love to do an oval head with just those eyes. Nothing else, just the gray green sad eyes.

Before sitting, she indicated to Bradovich that he had better lower the shade over the kitchen window, and he did so, realizing that this would only pique Mrs. Kastner's curiosity more than usual. The old woman sat quietly while he boiled water for tea, Earl Grey—he showed it to her before placing the bags in the teapot and she nodded affirmatively and smiled (bathtub white dentures)—then placed two cups and saucers on the table. He removed a box of butter cookies from the cupboard and showed it to her, but she mouthed a no and he returned it to the shelf.

Shortly, they sat facing one another as they sipped tea. She became warm and undid the top buttons of her cardigan. From her wrinkled throat hung a gold pendant that glittered under the bright light. As she raised and lowered the cup of tea, the flickering pendant's reflected gold light struck Bradovich's eyes and he felt a pleasantness slowly overcome him, a warmth permeating his body, and a love of sorts for this lovely old woman who hadn't yet told him what it was she'd come to warn him about. That she was a warning, he'd had no doubt from the very beginning.

The pendant glittered. He thought he would close his eyes for a mere second. Did so. Shuddered awake. This is nuts. The pendant glittered. Closed his eyes again. Abruptly came awake. Cut it out! Closed his eyes. The road was long and straight as a rule, and the sun was against him, and he was very tired. Wake up, you fool, wake up! The old woman was standing over him with a snub-nosed black revolver at his temple. He pushed her away roughly, started to rise, his lips taut from the onrush of blood. She quickly raised the slate to his eyes. On it was written, "You fool! Trust no one—not even your best friends. Incidentally, your apartment's bugged."

He nodded, yes, he knew.

She tore off a paper towel from the roll on the counter nearby—she moved now with great energy and speed—and wiped the words clean from the slate, then wrote in a firm hand, "Do you trust yourself?"

He took the slate from her, himself wrote, "Who the hell sent you to frighten me?"

"One does what one must."

"Why?" he wrote.

Now she stared long into his eyes, reading without emotion. Her sad look had been replaced by an impatient curiosity.

What the hell did she want from him?

"You're arrogant," she wrote.

He wrote in answer, "Of course. I enjoy living."

"You mean, " she wrote, "no one can live without arrogance?"

"At least some arrogance, " he wrote. "Otherwise we just drag our asses along."

She read his eyes again, and he didn't turn them away. For every sin he'd committed, he'd done penance. He hadn't needed a sermon.

"You asked why before," she wrote.

"Yes, why?"

Having read to the end, question mark, she wiped the slate clean, wrote, "That's the most important question. Think about it."

"I have thought about it," he wrote. "In fact, I think about it every moment of the day and night. Awake, asleep, it doesn't matter. I think about it even when I'm unaware that I'm thinking about it. It's always with me—like the shadows The Authority has sent to tail me everywhere. You've avoided my question, I repeat it: Why?"

Slowly, with a nervous hand she erased the slate, fidgeted with the chalk for a good five minutes while Bradovich sat impassively before her, then with a tic of her cheek, and then another, she raised the slate and quickly wrote, "You killed her, didn't you?"

Wizard didn't have to answer that. He'd broken his promise to Val. She lay in their bed in the bedroom under a lazy light and told him the time had come, she couldn't endure another night, the pain was just too great. He'd promised. "Do it! Now!" Yes, he had promised, but he couldn't do it. She looked at him with pure hatred, then screamed all night, and then all day into a second night. Called him a coward. Called him a goddamned Hunkie Jew son-of-a-bitch. The pain was destroying Val right in front of his eyes. It was hardly her anymore. It wasn't only her body which was being destroyed before his eyes, it was Valerian, his wife, that proud, dignified human being he loved more than his own life. Yes, he was a coward. Couldn't do it. All it would take was a double dose of morphine in the syringe. "Do it!" she screamed. Do it! he told himself.

She was just a husk. Bones and skin, except for that fucking plague in her belly, that extended ugly belly. But she was still Val. That second night she stopped screaming, she contained her pain, she was a tough broad. He sat at the side of the bed under the dim light. He'd taken all she could give, and all he could give himself. Her eyes were sunk deep in her skull. That's all it was now, just a skull, her once thick

111

brown blond hair now gray string, an old used floor mop. But still she was Val. She would be Val until she died. She would be Val so long as he lived. And Laura and Martin and their children, too.

"Please, my love," she whispered. "I lived my life with dignity, at least I tried. Let me die with dignity. I want you to remember me that way. The pain's just impossible, I want to be rid of it at last, want to be rid of it before I am no longer Val. Please, Wizard, please, my love."

He merely sat near her, holding her bony hand. "Do it while I'm awake, so the last thing I see is your face." She knew that at last he'd acceded, and would do it.

The bed lamp was on, that was all, and she stared at him through black holes in her skull. He leaned over her, kissed her, and gave her a triple dose. He sat at her side, holding her hands, and she smiled, her eyes fastened to his. And died that way. The doctor wrote, "Died of natural causes."

Bradovich didn't have to answer this strange old lady who sat before him, slate in hand. He'd answered to himself, and had been satisfied with his answer. He had loved Val as much as any human being could love another. He'd done what he had to do.

With an impatient shrug, the old lady again wrote, "They will toy with you, run you a wretched race until you concede to them and do their bidding. Kneel. Beware!"

He grabbed the slate from her, but before he could finish writing, "Nonsense—just plain nonsense!" she clasped the slate with two bony hands and pulled it from him, wiped it clean, rose and walked erect as a cadet, Bradovich following, to the closet where he fetched her coat and placed it about her slight shoulders. She raised a finger to her lips, then placed it gently against his—it was ice cold— turned and was let out without further ado.

112

He stood at the open door until he heard the clanking of the descending elevator. Who the hell was she, where'd she come from? Then Bradovich leaned with his back against the chained door and said aloud, "Yes, why?"

*I*n jaw-clenched fury Bra-
dovich paced back and forth through his apartment.

From his bedroom he could see the sharp-nosed Mrs.
Kastner, hair in pink curlers, spraying windex on her windows. She was rasping out a Fats Waller song.

> Don't let it bother you
> If skies are gray. . . .

She saw Bradovich staring at her and waved, then as was her
custom showed him the length of her scarlet tongue. Nut-
tier'n a fruit cake in San José.

Everyone knew he was in trouble with The Authority, but no one knew why—neither did he. He hoped and
prayed his children wouldn't hear of it. The sins of the father
should not bow the heads of the children.

He was no more guilty and no less innocent than the
next guy. The whole thing was a charade. He suddenly
remembered the charade trial in the park tunnel. Children

playing serious games. If we play the game, perhaps we can forestall reality.

Hares and hounds. Cat and mouse. What the hell did I do? What? He pounded heavy fist into strong palm. He paced faster. Ignored Mrs. Kastner's attention-attracting devices. The sun was out in full force, the days were becoming warmer. He ought to get out.

Without pulling down his blinds, he undressed quickly. Pulled on his black jogging suit, white sweat socks, jogging shoes, stretched a red sweat band around his head. Laughed aloud. The red and the black, the colors of anarchism. He did what he did. *"Pinga en su boca!"* he yelled so loudly the electronic bugs exploded the ears of his eavesdroppers. Mrs. Kastner who hadn't removed her eyes from him as he undressed and dressed shrieked with laughter and waved her skinny fist at him. He bowed to her. Rushed from the apartment, not forgetting to lock his door, but knowing it hardly mattered since they apparently had a key to every lock in the world.

Wizard began jogging as soon as he left the building. As he approached the oval in the park, he passed Floriana just leaving, her reddish tan cheeks wet with perspiration. They smiled at one another. Soon he was jogging smoothly around the track, a strong steady lope, his energy still fueled by a fury that could kill. He wouldn't yield to them, he would lead his life as he wished to lead it, as he had always led it. Go back to your studio, get to work. It's time. Why, why couldn't he go? What kept him?

He quickened the count. His feet pounded the cinders, he passed other joggers, his heart began to soar, his pores opened, the sweat flowed, faster and faster he ran, was an escaping prisoner, the bloodhounds snarled at his heels, a steel band girded his chest, tighter, tighter, faster, faster, stars exploded behind his eyes, and the steel band snapped and he could almost hear the sonic boom. He was free! He flew over the cinders as he never had since the days of his youth, running through the streets of Chicago, over its many bridges, to meet his friends in the park for a game. Now free, he could elude them. Round and round the oval he ran, Olympic gold medalist, outrunning the world, the galaxy, the universe. Time.

When Bradovich's anger had finally exhausted itself, he slowed down, fatigued but at peace. Now he walked and walked until his heart beat was normal, the pulse at his temples unfelt, and without thinking his feet led him toward the tunnel, a shortcut to the avenue, then he would stroll back over another path through the ball fields.

There was Cleopatra giving her sun god the finger, and there was the tunnel. It was partially lit by daylight flowing in from both sides, and he could see the benches on either side notched by many knives—initials, arrowstruck hearts, obscenities, the arch of the walls spray-painted with a motley group of figures in outlandish attire, faces white white.

He laughed. Of course. He'd hallucinated from fear of them. Them. It. The fucking, the cocksucking, the bloody, the stonefaced Authority.

Outside the other end of the tunnel, two small boys were kicking a soccer ball back and forth, and screeching with glee. He stopped to observe them. They were totally immersed in the sport, kicking the ball back and forth with great energy, each kick accompanied by a squeal of delight.

"Hey, kids," he yelled, "what about me?"

The boy with the ball didn't hesitate, but turned slightly and kicked the ball toward Bradovich, the Wizard himself. It came toward him in a lopsided roll, and before he kicked it Bradovich bent over to get a better look at it. It had an Apache topknot, a nose and eyes, and the face was still painted white white, scraped off in places where the boys had

116

kicked it. It was the head of the young woman whose fate he had judged.

He bent low to pick it up, as the boys wailed in disgust. Only when he felt its weight in his hands and stared into its wide open dead eyes did Bradovich scream from a terrible pain in his heart.

*B*ursting with importance and excitement, the small boys led the police and Bradovich to where they'd found the head.

It was a low knoll with junglelike bushes, their spiky limbs stabbing styrofoam cups, bits of newspaper, concealing in the remaining snow underneath pop bottles, beer cans, used condoms, a poster—SAFE SEX!—and a pair of woman's red panties. But no body.

Bradovich joined the police in a search for the headless corpse, having first given them a description of the trial in the tunnel at which he'd been coerced to act as judge. They never asked him to repeat it, and, observing their incredulity, he never volunteered. To them it sounded like a story fabricated by a crank who sought television celebrity.

And soon enough that was his fate, for as they beat about the underbrush searching for the young woman's body, they were joined by several television reporters and their cameramen, and Bradovich was forced to repeat the story of how he'd found the head numerous times, having been asked numerous, often stupid questions.

After several hours of hard labor in the wilds of the park, the body still not found, Bradovich left.

The telephone was ringing when he entered the apartment. It was Laura and Martin who'd heard the report on the early news and then seen him on television. They'd come together so they could talk to him. He related the story about finding the head as dryly as he could, and then Laura, who was not the subtlest of people, said, "You didn't look so hot, Pop. Tired. Drawn. Why don't you come here and get some sunshine and family loving." He assured her he was all right and would come as soon as his personal affairs permitted. Martin, less excitable than Laura, hoped his father was in good health and would soon find it possible to visit. "Alexander keeps asking for his grandfather." As Bradovich hung up the phone, he was glad neither had offered to come east to visit him.

Again the phone rang. This time it was Beatrice. In an excited voice, breathless almost, she told him she'd seen him on television, "You're photogenic, you know, in an ugly sort of way, especially since you've let that gray brush on top of your head grow longer."

Yes, he'd forgotten to go to the barber. Lately, he had other things to think about.

"How about coming over for lunch after you've cleaned up?"

"Sure, I'd love to." He realized her curiosity about the dead girl had excited her and overcome the hesitancy she'd lately revealed about appearing too aggressive. She *was* aggressive, and he liked her that way.

When Beatrice opened the door for him and he entered her flat, he smiled to himself. She was definitely pretty now, she simply shook with some inner vibrancy, and she glowed.

Curiosity had switched on her lamps. The rush of blood had flushed her cheeks so that there was a definitiveness to her features. Now I would know what to do with her face and head. Crop that red cloud so it became a French *coup*. It was a rare woman who knew how to reveal the beauty of her head with all the stupid hairdos one saw about. There was also a sensuousness about her that had not been discernible before, even when they'd made spontaneous love. It had been too dark in the hallway anyway. Still, it was only among human animals that one could find a vicarious pleasure derived from another's pain and suffering, or even from a violent death. He wondered if Beatrice realized how much this bizarre death had affected her?

"Stop staring at me as if you've never seen me before," she said, not neglecting to run her palm across his upper back. Yes, very pleasant indeed. One of the perks earned by sudden celebrity.

His jacket hung, his bird stirring in its nest, Beatrice led him into her dinette where the dishes were already set for two, and a serving platter of tuna salad with pignolia nuts and juicy green Israeli olives satisfied the eye. Whole wheat bread was in the toaster about ready to pop, and on the kitchen range the coffee pot was emitting a fragrance which comes only from fresh ground coffee beans.

No sooner had they begun to eat than Beatrice asked, "Tell me about her, the girl's head I mean. Was she pretty?"

"Yes, she was."

"What color were her eyes? They were open, weren't they?"

"Yes, they were wide open."

"Was there still blood where it had been chopped off?"

"Not that I could see."

Beatrice showed impatience with his short unadorned answers and pressed him for every detail, the gorier the better, and Bradovich decided to test the depth of her vicarious lewdness, so he began to expand and embroider upon the facts, concluding with the image of the young woman's lips being slightly parted, the tip of her tongue protruding, her dead eyes soft, a very beautiful young woman caught in the middle of the act of fellatio.

That the image was obviously invented did not matter to Beatrice. She sighed deeply, and closed her eyes to better savor the scene. The meat cleaver sweeping down in a murderous arch, Beatrice screened out.

Wizard smiled at her and thanked her for a delightful lunch. He then stood, as if preparing to leave, knowing damned well he wouldn't, and as he rose Beatrice rose with him and came at him in a rush. Their bodies collided, his arms swept around her slender waist, her arms around his neck, and they kissed violently. He had never seduced a woman in this way, but what the hell. He didn't stop to think that it was the other way around. She led him to her bedroom.

"A little slower this time, Wizard," she whispered, then smiled glowingly.

"I have no place else to go."

Beatrice had a comic lackserious imagination, and Bradovich was no slacker himself, and there was no rush, so they had a great circus of their own making. Three hours later, he left Beatrice assuring her he would come calling soon.

As he floated airily, a smug smile on his tough face, down the long corridor to take the elevator to his floor, Bradovich remembered that he hadn't thought about The Authority all the time he was in Beatrice's exhilarating bed. Oh, yes, the bird had soared. Perhaps he ought to move in with her, he was pretty sure she would welcome him.

Wizard didn't soar very long.

Though his outer door was locked as he'd left it, nailed to the door of his bedroom was a note typewritten in gothic letters on heavy expensive bond:

SOON YOU WILL HAVE TO ANSWER FOR YOURSELF!
PAIN GIVEN IS PAIN RECEIVED.
PREPARE FOR WHATEVER!

What pain? Who? Where? For the petty lies? The stolen caresses? The mercy he'd shown Valerian, his wife? He lifted his hand to rip the sheet from its nail, but at the last moment decided not to. Let them come and see that their warning couldn't make him tremble. He stretched out his two hands before him. Not a tremor. But he wished the time for decision was at hand, and that he could face his accusers whoever they were. If I have to stand on the dock, why wait, I'm prepared. To their whatever, I reply up yours!

How unsubtle they were. Simple-minded, perhaps.

To be sure, he wasn't very subtle himself. Could it be that The Authority had only recently instituted itself and therefore its operatives were still amateurs? You could almost immediately see the difference between a seasoned pro and even a topflight rookie, between the hands of an experienced master and those of a precocious student. No, no, his troubles were making him egocentric. Since he wasn't a subtle man, they were not finding it necessary to treat him subtly. The Authority was as old as history but had now expanded the number of people accused with doing whatever it was they did, and so it had had to enlist new though less experienced operatives. It could also be that The Authority's power was so overwhelming it had no need for subtlety, felt no need for tact, though it did seem to follow some laws, laws of its own making, spontaneously changeable, and totally arbitrary.

Even dead, just a head without a body, topknot and all, white paint flaking from her waxen cheeks, the wide-open staring eyes had looked unafraid, dignified, challenging even. She had never conceded, and perhaps that's why they had taken her life. But damn it, I'll stand with her. Yes, I'll stand with her!

To distract his mind, he switched on the television set. As soon as it brightened, the screen was hit with a blizzard of snow. He rose to reset the channel but before he could touch the dial it cleared itself, and on the screen appeared a faceless form, square-shouldered, unisex, who spoke in a metallic voice, evenly, without expression.

"You will be given ample opportunity to take the stand. For your own sake, do not hurry it. Surveillance continues. You are confined to the city limits. However!"

As the form disappeared from the television screen, Bradovich, still tongue-tied, expressed his frustration with a series of grunts, and finally with a howl of rage. I'm becoming an animal. The regular television program had resumed, followed almost immediately by a frenetic commercial. He couldn't bear the sound of the screeching voices, so he switched off the set and raced out of the apartment. The boulevard would give him comfort. He hoped.

And so it was. As soon as he found himself in step and in tune with the berserk bedlam of the early evening goings-on north and south and crosstown, he relaxed. Here life breathed, expanded, burst. The mighty throb of the streets revitalized him. If it had been but a few months ago, its pulse would have sent him running to his studio to work on a new block of stone, virgin in its uprightness, but not now. His imagination was as dried out as a fossil. He'd worked too hard for too many years, and after Valerian's death twice as hard. He needed a rest, a vacation. Yes, so here he was in trouble with The Authority. Forget it! Enjoy the street! What an oriental bazaar. Every color of skin, every shape of eye, of nose. The firm and infirm. Look at those heads. Round heads. Long heads. Some square at the chin, flat at the pate. Others long on top, almost chinless. Flat heads, round heads, squashed heads, elongated heads. And the chic, the laughing, the fresh-skinned women clothed from Bonnard's palette, how they took his breath away. My God, how I love them. The longer his trouble continued, the more he noticed a heightened sensuality. How he'd enjoyed Beatrice's body just a few hours ago. How supple she still was at her age.

"We're doing okay, aren't we, Wizard?" she'd cried out. "At our age, too!"

"We still have a hundred years to go, kiddo," he'd grunted, breathless.

They'd laughed and laughed afterward. And now he felt he was ready for more. The more you have the more you want. Enough is never enough. Like money. Like happiness. And for some people sadness. Bradovich had an almost unbearable desire to touch and feel every woman who came near him. Right at that moment he began slowly to follow three young women in purple and cinnamon and orange and maroon cotton scarves, their shanks in leg warmers, who strode before him with the arrogant grace of trained dancers. He wanted to embrace all three. To feel their smooth skin, their firm buttocks and breasts. He wanted suddenly to shout, "Live and let live!" to the crowded streets. And so he did. "Live and let live!" he shouted, and there was an echoing cry, followed by loud laughter. "Live and let live!" the street shouted.

At the corner, a young Asian girl stood playing a violin. Dvorak—from *Humoresque.* Not smoothly, but with great vigor and feeling. When the young fiddler finished the movement, she bowed low to her listeners. Bradovich and the others who had stopped to listen applauded. Wizard threw a half dollar into the young woman's open violin case.

At another corner, a black man dressed in somber black head to toe was speaking to a crowd gathered about him. His voice was husky, melodious, his diction that of a man of education. As Bradovich approached, the orator was in full voice. The boulevard stopped in mid-step to listen.

". . . are murderers, the most vicious species nature has evolved. By accident. All by accident. We are an accident, accept it. If there had truly been a whimper instead of a big bang, we would merely have been motes of dust whirling indiscriminately through space. And perhaps we are, yes, perhaps we are. But we are bad luck. Every species receives what it gives. Pain for pain. Joy for joy. Brutality for brutality. We all deserve each other. We are mean, shallow, avaricious, and, believe it or not, generous, intelligent, clever, wiseasses. Who of us cannot be bought? Penny for penny. Buck for buck.

The cockroaches in our kitchens laugh at us. Let them laugh. When we're gone, they will start on each other, have city-wide wars. What luck. Ha ha ha ha."

"So what's the sense?" a man in a baseball cap shouted. "Let's go out and fuck the world."

"Aaah men," a woman sang out.

The crowd snickered.

The Black continued to speak, but no one listened now. His magic had been dispelled, they all moved on. Another doomsayer. Wizard just stood, pondering the universe. Is there any sense to it?

From a stereo and video store a sensuous voice was heard singing,

> Everybody get upset when a bat fly in me face,
> Fly in me face, tonight. . . .
> . . . when a bat fly in me face,
> Fly in me face, tonight. . . .

A reflection in a shop window informed Bradovich he was being followed by a new shadow this time. Short, sturdy. A young one. Muscular, but a real amateur. Bradovich swung from his heels and caught the operative dead in his tracks, struck him in his solar plexus. Disjointed his nexus, the little prick. The operative doubled over and Bradovich slammed a knee hard to the crotch. As the operative groveled, Bradovich axed him with a stiff hand to the back of the neck. He who gives gets in equal portion.

No one seemed to notice. Bradovich strolled on, whistling the fiddler's tune from *Humoresque.*

*T*he entire building was overrun by police.

There had been a break into Floriana's apartment. She had returned from a run to find her rooms being ransacked by two masked men. Consuela, her young maid, was half naked and tied to a bedpost. She'd been flogged. Dauntless, Floriana had drawn the knife from its sheath on her thigh, but the invaders were too quick and disarmed her. She fought them off, ripping at them with her fingernails, but they beat her down and then raped her, after which they taped her mouth and bound her.

Before leaving, one of them had said, "That'll teach you, baby. You're in trouble."

They had then fled to the roof, sawed off the huge padlock that held the thick metal door, and skipped across the roof to the next building, a hotel for elderly pensioners, descended by elevator to the lobby, and calmly strolled out. A client who had come for a matinée appointment with

Floriana found her an hour later. He released both Floriana and Consuela from their bonds, then called the police and a doctor.

Bradovich sped to her apartment and arrived just as Floriana was leaving with her doctor, an old and beloved client, to go to the hospital to be scraped clean of the unwelcome seed. Her face was bruised, her left eye swollen and turning black and blue.

They embraced, and she muttered in his ear, "I'll get those bastards if it's the last thing I do."

"Good for you, Flor, I'll be with you all the way."

Before departing with the doctor, Floriana asked Wizard to phone Golo before he got a garbled sensational report of the attack by radio or television. What troubled her the most was the parting remark made by one of the hoodlums.

When Bradovich reached Golo, the Gimp became hysterical. After being calmed by Wizard's ice-cold voice, the Gimp finally calmed down and said he would go immediately to the hospital to see Floriana.

The phone was ringing as Bradovich entered the door of his own apartment. It was Beatrice. She was excited, curious, in a hurry to see him. "Is Floriana in the same sort of trouble you are?"

"I'm not in trouble," Bradovich said coldly, and with an emphasis which belied the denial. "As for Floriana, it was the usual break-in and assault that occurs every day in the city, and now—" Someone was breathing down Bradovich's neck.

He spun around, and stood chest to chest with one of his shadows. All he could say was, "Son-of-a-bitch!"

"What? What?" Beatrice shouted.

"Beatrice, darling, I'll call you later tonight. Yes, yes, I'll come for a visit. Right now I have an unwanted houseguest. So calm down, honey, I'm all right."

128

Stoneface merely handed him an envelope, did a martial about-face and left as he had entered, soundlessly. This one was a pro.

Heavy, expensive bond paper, the type gothic:

"What occurred at Floriana's was unforgivable. Those responsible shall be severely punished. You may inform the Gimp of that fact. As for you, Mr. B., you will be given ample opportunity—or have we already said that? Albeit."

Bradovich had the distinct impression that The Authority was slipping. The corruption that flows from total power is so pervasive it begins to debase power itself. What follows is ineptness, slackness, laziness, contempt for power. Soon openness followed by revolt. Bradovich laughed at his optimism.

He phoned Golo to give the Gimp The Authority's message, but as he had expected Golo had already left on the run. So he called the information desk at the hospital, and there caught the Gimp just as he had received Floriana's room number. "They owe me one, that's certain," Golo said, then hung up to rush to his beloved.

Wizard just sat in the semi-darkness of his dinette for an hour. He felt nauseous, sick. Was his ability to fight them off so far turning them toward Floriana? Was his good fortune to become her bad? You're becoming superstitious, Bradovich. What happened to Flor had nothing to do with you. It was separate, apart, and without connection. Yeah, arbitrary. The word arbitrary set Bradovich's fear off on a wild chase. He began to sweat, to feel light-headed, tremulous. He tried his old routine of the night before a big game. Breathed slowly, one and two, one and two, one and two. It took some time, but slowly it began to work. He lay on the hard kitchen floor. Felt better. Slept for a half hour. Awoke. Was hungry.

Showered, shaved, aftershave splashed on, he dressed in fresh clothing. He was going to see his girl. Grinned. Yes,

129

that's the way he felt about Beatrice. Like a young kid. His girl. Why not? They had that feeling about each other. As excitable, as uncool, as she was, so unlike Val— Don't start that crap. She's she and Val's Val. Yes, they had a heart-jumping feeling about each other. Why not? That so far hasn't changed. Infatuation. Love. Kids become infatuated. Adults feel love. Or was he kidding himself? Whatever it was, their blood quickened when they spoke to or saw each other. Strange what got her sexually excited, though. Everyone to his or her own. Beatrice's head, despite that silly hairdo, moved him, excited him. Made the bird raise his old head. Ha. Enough of that, let's eat.

Downstairs, Slim wanted to talk to him about what had happened to Floriana. Whenever Slim mentioned her, he smirked, and it made Bradovich impatient. This time he didn't even listen, but walked past the man. Fool. He went to a Thai restaurant in the neighborhood where the exceedingly spicy food cleansed the funk out of his mouth and intestinal tract. As he left, he noticed he was being shadowed by a man and woman team. If the she got too close, he wouldn't hesitate to rap her one. He was an equal opportunity employer.

Before going to Beatrice's, he decided to stop in to see Floriana who had returned from the hospital with Golo. Her luxurious apartment had been put in order by Consuela, none the worse for her flogging, her curvaceous body a replica of Floriana's but in smaller dimensions. "Not too bad," she said with a modest smile, "I recover quickly."

Golo was already gone when Wizard arrived, having, he had informed Floriana, an engagement at the London Palladium beginning the following day. "Lots of long green," he had told his *amie*.

Flor was in good spirits. Her bruised black eye added another facet to her beauty, a piquancy of sorts. *Salada* in her language. Spicy. The Gimp had assured her that he would see to it that she would no longer be harassed by whomever it was. She had total trust in him and believed every word he said. A mistake, of course, always a mistake, as Wizard would have told her had he the guts.

"Perhaps I'll do as Golo wants," she said to Bradovich, "give up my business and marry him. It's time I gave up my adolescent attachment to my brother. Enough. Golo's such a tender man." She closed her almond-shaped eyes, wet her lips with the tip of her tongue, then smiled secretly.

"What a pleasant surprise," Bradovich said. "Never thought you'd give up your business. I'm happy for both of you."

"Let's drink to it," she said, smiling to reveal her uneven white teeth and red tip of tongue. Aside from the bruised eye, she showed no other ill effects from her brutalization earlier.

As she, Consuela, and Bradovich sipped Napoleon brandy, Floriana told him that Golo had convinced her to give up the knife for a pistol. He himself was a crack shot and would give her instruction in its use. (Why don't I get a gun? Bradovich wondered. Though he had the feeling that if he were to put a bullet through one of his stonefaced companions, it would be like putting a bullet through a ghost.) Until the gun was in hand, Floriana said, she'd decided to sheath her knife in the holster strapped to her thigh. To prove it to him—as intelligent as Flor was, as insightful, she could never prevent herself from being a coquette and would surely drive Golo nuts for the rest of his life—she raised her pleated woolen skirt to her hourglass waist, and he saw the sheathed dagger as well as her exquisitely groomed orchid. Wizard's nose as well as his penis twitched. If he hadn't had a date with Beatrice for that very night, he'd have been tempted to ask his friend for a favor despite Golo. "A stiff cock always

131

takes precedence over friendship," the Gimp himself had once said.

Instead, he told Floriana and Consuela that Beatrice Holden and he had at last found each other.

Leaping to her feet, Floriana rushed to his side, bent over as he sat and kissed him on the lips. "About time, you old fool. Valerian, I'm sure, is happy for you. You're lucky, Bea's a kind woman. And good for her, too, you're a nice guy, just a little nuts, that's all. But so is she. A real good match. She's tall like you, big where she has to be—a good match, yes, indeed."

"We are all *loco, muy loco,*" Consuela said with a loud laugh. The brandy had made her slightly tipsy, and blood flushing her face had turned the olive of her skin to a light mahogany. He had the right piece of wood to do her head, Bradovich thought.

Rising to his feet from the low chair, he said, "Floriana, don't worry, those two hoods are already being taken care of."

"Yes," she said, "Golo told me."

The two women saw him to the door, each kissing him before the door was closed and chained behind him.

As he slowly mounted the three flights to his apartment before going up two more to Beatrice's, he was praying that he wouldn't discover something amiss. He just wasn't up to it at this point. As he was about to insert his key in the lock, the door to the next apartment opened, and his neighbor, Mrs. Kastner, came reeling out, weeping bitterly. She lunged at him, and before he could utter a word threw her skinny arms around his neck, placed her head against his chest and soaked him in her tears.

There was little he could do but just stand there, allowing her to cry on his chest, hoping it would end soon, for he was anxious to freshen himself and get up to Beatrice.

When Mrs. Kastner's sobbing began to abate, Bradovich at last asked her what was wrong.

"That so-called man I've been married to for twenty-five years, that alleged man I've loved and until this very moment I thought loved me," her sobbing increased again, her tears soaking through Bradovich's shirt to his skin. Again he waited patiently. And again she caught hold of herself, her sobbing lessened, and with hesitation, the river of words coursing over the obstacles laid down by her anguish, she finally told him that she'd come home from shopping, knowing her husband would be waiting for her, but instead of him she'd found a note on the kitchen table. "I didn't have the courage to face you," he'd written. "I've had enough. Be well. Forgive. Ira."

"Damned coward," Bradovich said. "A real louse. You're well rid of him."

"Don't speak ill of him," Mrs. Kastner snapped angrily, spun about on her worn carpet slippers, and re-entered her apartment. Not even a thank you for lending her his chest now soaked with her salty tears.

Fortunately, Bradovich found no new message from The Authority. The one nailed to his bedroom door remained untouched. He washed up, added a bit of aftershave, combed his longish gray brush, and changed his shirt. Decided to brush his teeth, too. He was going up to visit his girl, have some fun. As he turned out the lights and headed for the front door, the telephone rang. To hell with them, let it ring. He didn't however open the door. They got me by the balls, those bastards! It rang and rang. An emergency, who knows?

"Yes?"

It was neither The Authority nor one of his children. It was Mrs. Kastner, growling. "He's returned, the prick. You're right, he's a stupid coward, and louse is too good a word for him." She hung up so hard his eardrum pinged.

*A*s Bradovich stepped into his apartment at 6:30 the following morning, the City Coroner phoned.

Before that Wizard passed an alternately peaceful and exciting night with Beatrice Holden. Her apartment was dimly lit when he entered, and she was wearing a translucent peignor under which her nudity promised mysterious adventures. They embraced, kissed, gave each other a feel.

"Let's have a bite first," she said, allowing herself a momentary teasing grind.

"I've already eaten, but a little more never hurt me."

On the large round coffee table in the living room under a soft lamplight was an array of delicacies, baby shrimp, oysters on the half shell, fried calamari, squares of German pumpernickel, and a sturdy very dry white wine.

He had barely chewed and swallowed his first shrimp, before Beatrice, her hair a spreading cloud over her head, asked, "Aren't you going to tell me about what happened to Floriana?"

She allowed him another bite and a sip of wine, before asking again, "C'mon, Wiz, I want it from the horse's mouth."

He smiled. "I'll call Floriana, she'll come up and tell you herself."

"Okay," she laughed, recognizing the comedy of herself.

"You look lovely," he said, his eyes taking in the softness of her under the peignoir. She moved closer to him on the sofa, and he felt the soft warmth of her breasts. He kissed her, caressing her with his free hand, the other balancing an oyster on the half shell. She pulled one of his ears.

"Flor found these two hoods in her place," he began, "they were wrecking her apartment. Floriana withdrew her knife from its sheath and went for them."

"My God, what guts."

"Flor's not a tame little pussy cat."

Bradovich sucked in another oyster, followed it with a sip of white wine. Paused.

"And?" Beatrice urged.

He took another sip of wine. Paused again.

"And?"

"Consuela, incidentally, was tied to a bedpost in her bedroom, half naked, flogged. Not too badly hurt, though." He paused again. Her cheeks were flushed, her lips swollen.

"And?" she whispered hoarsely.

"And then," Bradovich continued, detailing the events that followed.

Beatrice was helpless now. "I can't move from here," she said huskily, as Wizard stood.

She lay revealed on the rug, a soft unbreakable

135

woman, observing him as he slowly undressed. "You still have lots of muscle, Wiz," she whispered.

No circus tricks that night. They spoke in whispers. They weren't kids. It was a long, slow night.

At the door in the pre-dawn hours, Beatrice told him she loved him.

"But I can't keep on telling you these violent dirty stories."

"You can always make up fake ones. It's only between you and me."

"Sure," he said, giving her one last caress under the arch. "As long as you know the difference between fantasy and reality, and keep each in its proper place."

Bradovich was wanted at the city morgue to make an identification. "Take a cab, we'll reimburse you."

The cab rolled quietly and swiftly at that pre-traffic hour.

At the morgue, he was ushered into the refrigerated room with its slabs for the dead. There was a strong stench of formaldehyde. He hated the smell, it reminded him of his father, Uncle Walt, his mother, Val. The last smell of them. The one you don't ever forget. The young girl's preserved head with its Apache topknot rested under a white light. The blue gray eyes stared. Her bloodless lips were parted to reveal tiny, perfectly shaped teeth. The tip of her tongue protruding from the left of her mouth looked like the streamlined head of a white baby snake. The shoes of the little boys who had played soccer with her head had done little damage to her marble cheeks now scrubbed clean of the white paint. She had the skin of a child, a beautiful woman child. Flawless white skin under the bleak white light glistened. At the

136

point where the head had been severed from the throat strings of cartilege curled like strands of sodden noodles.

The mortician and an aide pulled out a drawer from the refrigerator for the dead and raised from it a headless body which they carried to the slab and placed beneath the girl's head. Though deflated somewhat, the body seemed to be proportionately right for the size of the head. The coroner assured Bradovich it was the body which belonged to the head, no doubt about it. He was a tall distinguished looking man, regal in his posture. In a movie he would have been consort to the queen.

"The little twit was sixteen years old," he said without emotion.

"How can you tell—by the teeth?"

"Don't be silly."

The body had a long gash down the center of its abdomen.

"What's that?" Bradovich asked.

"The heart, kidneys, bladder and liver have been removed," the consort of death said.

"Why'd you do that?"

"We didn't do it. It was done by whoever beheaded the little cunt. At least, we assume that. The body, incidentally, was found by a policeman near the area where you discovered the head, at the foot of the obelisk. The wind had blown leaves, underbrush, street garbage and the like off it. Scum bags all over the place. Can you identify her?"

"Of course not. Why should they do something as crazy as that, cut out her vital organs?"

"They're probably in the business of selling body parts, like car thieves. We need an identification."

"I saw her only once, and under a bad light. And dressed—more or less. Oh, yes, her breasts protruded over her bodice or whatever. Small, pointy, like these."

137

"That should be enough identification."

"Why? Half the teenage girls of the world have small, pointy breasts. Maybe there's another head some place. Who the hell knows what's going on out there?"

"That body and that head go together. No doubt about that."

Bradovich had the feeling he was talking to no one.

"Hasn't her family come forward?"

"A picture of her head has been broadcast on television and by poster. No one has stepped up to identify it so far. And she doesn't fit the description of any of the missing persons so far reported."

"She certainly was a pretty kid, wasn't she?"

"Yeah, a spicy little piece. I wouldn't have minded diddling her myself. She was no virgin by a long shot, or a small one, for that matter." He laughed. "A man can get excited just cupping one of those pointed tits."

He was almost salivating, and Bradovich wanted to punch him in the nose.

"She had great dignity, and much courage," Bradovich said. "She didn't look like the kind who'd stand for diddling."

"Probably a runaway from Montana or Salt Lake City. This town's full of that kind of gash."

"Sew her together and bury her, you prick. If necessary, I'll pay for expenses."

The consort of the queen took in Bradovich's size. When one was that big, one could say almost anything one wanted. "They fucked her in the ass, you know."

"Against her will, no doubt. She was not the kind who liked to be buggered. A tough kid. Dignity, with great dignity. . . ." He was talking to himself again. This time the coroner and his aide had left him.

As Wizard strode angrily toward the lobby of the city morgue, he glimpsed into the far reaches of a dimly lit

138

corridor, marble-walled, marble-floored, and saw two white-clad figures in a violent amorous clinch. He stopped, stepped back the better to observe them, saw them fall to the floor and really go at it.

Love and death—or maybe it's the formaldehyde, he thought, as he left the marble halls.

Two tall, square-shouldered stonefaced men dressed in black fell in behind Bradovich as he, having decided that the long walk would do him some good, turned uptown.

Wherever Bradovich went, he sensed that all about him were signs, indications, whispers, "Beware!" He was not a cautious man. Had never been a cautious man—as a kid, as a ball player, as an artist. Caution inhibited energy, and energy was his strength, his forte as the critics said.

But Bradovich was no fool. He knew there were times when even someone like himself had to conserve, preserve, even cosset his energy. And now was that time. Just ignore them, ignore them with the hope they don't touch you, that's all. He didn't like strangers, though by now these shadowy yet substantial figures were hardly strangers, to get physical with him, to push him around. Whatever. However. Wherever. Since the concrete cornice had missed him by inches, they'd not tried anything rough. If they kept their distance, he'd keep his.

That was the decision he made. He would accept the surveillance; further, ignore it, just so long as they kept their hands to themselves. The Authority wants to expend its energy to have him followed wherever he went, even enter his apartment to leave notes, *et cetera*, okay, go ahead, be my guest.

Sure enough, decision seized, Golo phoned.

Because of the crude manner with which Floriana had been treated, The Authority, as a favor to Golo the Gimp, to lift his spirits, so to speak, had granted his friend B. a stay. The investigation and surveillance in re: Bradovich was estopped, pending.

"Pending what?" Bradovich asked.

Silence, then, "Be satisfied. You have a stay."

"You're right. I'm just greedy. Thanks a million."

"Go visit your kids. Marry Beatrice. Take advantage of it when you have the edge."

"You're right, you're right. I'll book a flight today for the West Coast. See my children, my grandchildren. Hey, Golo, you've saved my life."

"Just a stay, bud. When you get to the coast, call my brother Ben. He's a ne'er-do-well, but a nice guy. I love him. He never asks for too much, just enough to keep him going. He lives in a doll house our parents gave us when they kicked us out of the family home because they were embarrassed by our shortcomings. Ha ha ha. He's a dwarf. We're twins, even though I'm a foot taller. My parents were straight, ordinary, normal size like all the rest of you geeks." He laughed with a growl.

"Sure, I'll be happy to look him up."

Golo gave Bradovich his brother's phone number, gave him his blessing, told him to be back in time for the wedding, Floriana and he would be mad if he didn't come. They said good-bye.

The day was gorgeous, the sun high, the sky Caribbean blue. The boulevard had been swept clean. Not a speck of dirty snow left on the center island. The bushes had been pruned, the grass cut. Not a styrofoam cup in sight. People walked with purpose. Car honking was at a minimum. Joggers had greater spring in their legs. Kids rode their bikes with terrific speed. Life coursed briskly through the city's arteries. Everyone was polite. Smiled. Civilized behavior reigned.

Wizard strode swiftly to Cosmic. The handsome travel agent with the whitest teeth on the boulevard smiled, heaved her bosom. Booked him a flight for the next morning, return passage three weeks hence. He bought a bouquet of roses, long-stemmed, of course, sprinted all the way to Beatrice's door.

"How lovely," she said, taking the flowers with one hand, reaching under his jacket to squeeze with the other. She hopped about with great energy, with sexuality, with love. He couldn't quite figure out whether she made him feel older or younger.

He hugged her to himself. Beatrice was a slender sloop, full at bow and stern.

"And how great for you," she continued, cupping what she had squeezed.

"You know?"

"Of course, everyone knows. Despite what people say, a good word gets around as quickly as a bad. Remove your jacket, take me in your strong arms. Let's copulate!"

Outside the apartment building, hardly discernible to the unsuspecting eye, lurked two slender figures, wearing tan head to toe. Later, coming and going, Bradovich, because he did not wish to see them, did not see them.

And so it was when Bradovich deplaned three thousand miles west. They were there, two slender figures in tan, lost in the besmogged airport.

Martin, tall, dark, and cool, and Laura, tall, blond, and excited, greeted Bradovich. Carlotta and Alexander screeched with delight, dancing around their grandfather's heavy legs. They formed a circle, toddlers in the center, and kissed and embraced simultaneously.

"You two look great."

"You feel hard as a rock, Dad," Martin said, smiled.

"Yeah. Jog regularly. Take long walks. Heft big blocks of granite."

"Pale, though," Laura said, always the most observant.

"I'll get a tan here if the smog lets up. You still wear your hair in that—"

"You just arrived, Dad. Ready to start a fight?"

143

"You have to watch the sun, Pop, even with all this smog. We have a first class sun screen in our line." Martin was the new type of salesman, cool, disinterested. He had you buying before you knew you even wanted it. Had been that way at the age of three. He was a good son, quiet, calm. There was lots of love flowing both ways.

Laura drove them to Martin's house in a big Volvo station wagon. He would visit with Martin and his family ten days, then with Laura and hers for ten more. On the twenty-first he'd return home. Somewhere in between he'd have to pay a call on Golo's brother Ben.

Martin's house was an eagle nest perched on a crag overlooking the Pacific. In his driveway lolled a Mercedes sports car and a BMW sedan. Not bad for the grandson of a WPA handyman and the daughter of a Kosher slaughter-house butcher. Martin, as big as his father, though without the old man's bulk, was a jobber. He jobbed anything from peanuts to yachts. He knew all the angles, and took advantage of every one of them. He couldn't help himself, he did it without even thinking. An honest man who played and won at poker without concealing aces up his sleeve. He had a knack, a talent for it. The house was what Bradovich called an in-and-outer—you lived inside and outside simultaneously. It was planted with palm trees and poinsettias, and surrounded by stone retaining walls, otherwise it would long ago have slid into the sea. At one end of the wall rested a head of Martin that his father had done when the boy was fourteen. "Sophisticated primitive," Wizard called it. Even at that early age, Martin had had a face almost as craggy as his father's. An early work, it showed little relation to the later polished abstract heads. When Bradovich saw it again as he entered the patio, he exclaimed, "Hey! that's a good piece, I like it." It made him feel so good, he went to Martin and gave him a bear hug.

144

Pegeen, Martin's wife, had come to greet them. She laughed. "What about me?" Bradovich looked at her. She was a slender bottom-heavy woman with a long pensive face, intelligent, sensitive. A shy woman, she stammered when she first met people, and it took her several encounters before she was able to relax. But Bradovich she approached with a big white-toothed smile. "It's been a long time," she said quietly. They embraced and kissed on the lips. Wizard liked her. She kept Martin, who like his father was an ogler of women, who yearned to conquer every one of them, in line. She smelled of lavender.

"Gonzalez, Pop."

Laura, feisty, combative Laura, her long gold Valerian hair concealing her face and thus its great strength and beauty—Bradovich had never hesitated to tell her it would better reveal her beauty to sweep her hair back tightly, thus provoking angry tears and argument—stood with her husband, a slight wiry man, his forearms in the short-sleeved shirt like entwined cables. He had a dark brooding face, alert smart eyes. The younger generation may not be wiser than their parents, but my God how much more intelligent they look. It must be TV that does it. Gonzalez smiled at his father-in-law without opening his lips. Seemed nervous. He and Laura had met, wooed, married, and had a child without Bradovich's presence at any time. Wizard had never realized it could happen that way until it happened to him. Children become adults no longer need their parents—had he needed his? When his kids were gone, he always missed them, wished they were around. After they were with him several days, he was glad they were independent and soon left. Like their mother and father, they needed lots of room in which to throw their elbows. Gonzalez gave the impression he was one of them. He taught American literature at the same state college where Laura taught philosophy.

On the terrace, under a sun in the cloudless blue sky—the smog was to the east and in the valley below—Alexander and Carlotta climbed on their grandfather's large lap, over his back, under his legs. Carlotta, with her raven hair and pitch-black eyes, looked as if soon she would be slinking into a room, a red rose between her teeth. Alexander, both his parents brunettes, had gold hair, and with his little stalwart body already looked like a Greek god. Of course, it was their grandfather observing them. He already had committed their heads to memory, and knew exactly the shape the marble would take. On occasion he would grab one, hug him or her, then send the child off with a pat. They kept insisting on touching him as if to get to know him better that way, and to him every touch from them was a moment of bliss.

Bradovich's experience with The Authority seemed to have left him without telltale scars, since his children, especially observant Laura, saw nothing on his face to inform them that he had been in any sort of trouble. (He pretended to himself that the stay was an acquittal of sorts. Fear makes one foolish.) They told him he looked well, had lost some weight since they'd last seen him.

Pegeen and Laura were in the kitchen preparing a late lunch. Martin and Gonzalez sat nearby, the latter puttering around with his pipe, which kept falling to the terrace. Bradovich, though a large man, never thought of himself as prepossessing, but for some reason his presence seemed to discomfit his son-in-law, and nothing he could say put the man at his ease. Martin was telling them about his business, which was obviously thriving.

Bradovich mentioned Martin's oft-repeated boast when he was still in high school that he would be a millionaire before he was thirty. He was now thirty-three.

"I made it easy, Dad." He said it simply, stating a fact.

To put Gonzalez, an untenured poorly paid assistant professor, at his ease, Wizard made the usual remark about money not being everything.

It didn't help. Gonzalez dropped his pipe. As he rose from picking it up, Bradovich remarked that he had an interesting head, spatially perfect between the temples, in balance with the curve from the back of the head to the nape. Gonzalez listened wide-eyed, so much attention to his head. Martin smiled to himself—his father's old trick with shy people. Who doesn't love to hear about his undistinguished head? Bradovich had his small pad and charcoal pencil in hand—he was never without them—and was sketching Gonzalez's skull and pointing out the axis and spatial points to his son-in-law whose pipe was now resting at ease on the terrace table.

At lunch they were all relaxed, laughed a lot. Laura was a good mimic and regaled them with jokes about friends. It was obvious to Bradovich that Laura and her husband were happy with each other, and that the two couples had established an easygoing relationship. Laura and Martin—she older by some eighteen months (Valerian, getting on in years, had wanted her children quick)—had always been close, though combative and competitive, stopping short of drawing blood. That they all had their problems, he had no doubt, but were not about to reveal them now. Alexander and Carlotta were healthy, energetic and looked smart, so what else was there to think about?

After the men, including Bradovich, cleaned up the kitchen, he lay on the lawn and permitted his grandchildren to climb all over him, he was their new toy. It was so pleasant feeling their childish hands and feet and sweet-smelling bodies that he fell asleep.

Later, he said, "Jet lag." It wasn't that at all. It was the sudden release of tension, the freedom from constant sur-

147

veillance. There was no shadow, so far as he knew, except what the sun made. And the full moon, of course.

On the road at the end of the long winding driveway, two slender operatives in tan slacks and polo shirts, white sweat socks and tan sneakers, who were bicycling past, stopped for a moment to nod their respective well-groomed male and female look-alike heads in the direction of Martin and Pegeen's inside-outside house.

The days passed at a leisurely pace, thus swiftly. Bradovich was pleased to allow his children to chauffeur him about to see this and that, a museum here, a fine building there—a couple by Frank Lloyd Wright, so much imitated but never equalled—a great restaurant to satisfy one's most delicate taste. The gallery which showed his stuff and the contemporary arts museum which owned several of his wood heads he refused to visit. They spent several afternoons at the beach, where he divided his attentions between his grandchildren and the remarkable assemblage of gorgeous figures, female and male. (He was secretly pleased that Laura in a bikini was spectacular.) He had an itch for his charcoal and pad, but restrained it. He didn't care whether they went out or not, he could have rested in one house or the other— Laura's was an old Spanish hacienda with highly rubbed teak paneling, a restful refuge from the constant daytime sun and heat.

Bradovich would have been satisfied merely to observe them at their lives, the children crying, having tantrums, or sitting on his lap smelling of baby caca, climbing all over him, the adults arguing (they were an argumentative foursome), but it made them happy to entertain him, boring him on occasion with chamber of commerce claptrap about their weather, the unique beauty of their area of the country (a territorial disease), all of which he attended with amused pleasure, so he did as they asked without demurral. With unconscious intent, The Authority, the possible temporariness of the stay were all concealed deep in Bradovich's large gut, buried under the sheer happiness he awoke to and went to sleep with every day. They did shed a tear or two when they spoke of old times, when Valerian, their mother, was alive, but that also was the sort of sadness which in a strange way yielded pleasure. Not nostalgia, which Valerian used to call *papier-mâché*, but honest remembrance of someone loved and lost. And time, too. Lost.

As Wizard's visit approached its termination, he phoned Ben, Golo's brother, the dwarf, and made an appointment to meet him at his house. Ben's voice was a growl, similar to Golo's, a pleasant growl, for he seemed pleased at being introduced to a friend of his famous twin.

Ben met him on the street under a frayed palm tree. He had a large head on a tiny stringy body, wore floppy purple trousers, Mexican huaraches, a carmine sport shirt. His nose had the shape of a dehydrated banana. Only in doglike grin did he resemble Golo. Where Golo was about four feet six, Ben was less than four feet. A gnome, a troll. Behind him leaned the miniature house the Gimp had mentioned.

His voice rising from his heels, as low a basso as Bradovich had ever heard, Ben said, "A friend of my brother's an enemy of mine."

As Bradovich stepped back startled, Ben growled a long low laugh. "Just kidding," he said.

"Of course."

"You're awful big, but you'll just about make it," Ben said, leading the way.

Bending almost double, Wizard managed to enter with Ben into a clean, tiny place furnished with children's rattan furniture. The walls were hung with huge nude blow-ups of famous female movie stars who had in the beginnings of their careers posed for centerfolds or calendars. Some had done so in lascivious poses with partners, male and female—cunnilingus, fellatio, buggering, all in an unlubricious sterile sort of way. There was also a wall-to-wall and floor-to-ceiling glossy photograph of Ben standing close to the fat lady of the circus, mountainous alongside him, and on whom had been superimposed with a charcoal pencil two fat nipples and a black hairy twat. And there as real as day was John Wayne in cowboy clothes, Ben in dress-alikes, the latter pointing a huge pistol at Duke whose hands were raised in mock terror.

Ben motioned Bradovich to a full-sized chair, the only one in the room—Wizard felt as if he'd been stuffed into a stage trunk—and he sat down. On another wall were tacked enlarged stills from *Snow White and the Seven Dwarfs*—was Ben Dopey? In a corner squatted a color television set, which was on but with the sound turned down, and while Ben was in his tiny kitchen making them lemonade Bradovich watched a silent daytime serial.

"Have it on all the time," Ben growled, entering with two tall frosted glasses of lemonade. "Keeps me company. Hard to find company. Who wants to be seen with a dwarf? Nearly all my friends are dead. We die young, I think, I don't know, never kept track. I'm over fifty." Said proudly.

"Wouldn't have figured you to be over forty," Wizard said, incredulous, sipping the refreshing lemonade in this steambox of a house.

"Because I'm little. Little guys always look younger. Fat people, too. And niggers. It's the skinny people who look their age. Full of wrinkles."

For a few minutes they watched the television program. "The stepsister's the man's daughter," Ben said, referring to the soap opera they were observing, "and his second wife's daughter. Catch? She's a goddam bitch. But I'd like to throw her one. And so would her stepfather, but she keeps playing him for a sucker, cockteasing him. In the meantime,

150

his own daughter is falling in love with his wife's, her stepmother's, lover. It keeps going around in a real circle jerk." To him they were alive and true and he lived every moment of their lives with them. "What you doing on the coast?"

Bradovich told him.

"What you do for a livin'?"

"I'm a sculptor. Used to play pro football. Right now I ran out of steam, am doing nothing. You're in the movies, aren't you?"

"Me, I used to get bit parts. Once played with the Duke himself." He stared, and Wizard with him, at the blow-up on the wall, and they were silent as the dwarf's mind did a backflip to the days of his greatest glory. Bradovich stole a glance at him and saw his face become pink and his eyes take on a look of disbelief at the sight of himself sporting among the gods. Then he sadly shrugged his narrow shoulders and closed the vision out with a quick blink of the eyes.

"Golo's the lucky bastard, got all the talent. Lucky for me, too, now—gets me an occasional part, and pays all the bills. With money I can always get a dame. Nifty ones, too, in this town. They come from all over—from Maine to Baja. From the North Pole to the South Pole. Real beauts with boobs like mushmelons. When they hear I'm Golo the Gimp's brother, all of a sudden I get special attention, there ain't nuthin' they wouldn't do then. Dwarfs are always hot in the pants. My wang's as big as any man's. We can do it all the time. Just like niggers and Chinks."

"I'm so glad," Bradovich said icily, which Ben didn't seem to notice.

Ben laughed, pointing his canines at his guest.

"But I gotta good friend," he growled, "the best." Rising on his little bowlegs, he went to another room. He returned shortly with something cupped in his hand, a

151

plump white mouse it turned out. He must have fed it milk and honey, so plump was it. It ran up his thin little arm, ducked under his open collar, rummaged around his body, emerged from the short sleeve on the other arm. "I call her Bette Davis," he grinned, his fangs showing. "Once worked as an extra on one of her pictures and tried to get her to spread her legs, but she refused, the fool, old as she already was. I'd have given her the best fuck she'd ever had. Look, my sweetie's got pop eyes just like the other Bette."

True, the mouse had pinkish pop eyes and a fat pair of loins. Without trying hard, Bradovich did find that the mouse resembled the old movie star, who really had never been his style, she'd been a bit too shrill and hammy. Bradovich had to admit to himself he liked his movie stars cooler and right out of porno magazines. If you're going to daydream, you might as well go the whole hog.

Ben and Wizard watched the silent TV as the daughter and her stepmother's lover lay on a bed seemingly naked under the linen coverlet. As she ducked under the coverlet, Ben said, "Ah, you see, she's gonna give him head."

Bradovich said nothing, and they watched in silence.

"Bette," Ben growled between his canines a few minutes later, "owns me and I love it. We own each other. She holds on to me like hell with her little feet. Without me she croaks dead. Without her I go buggy. Just the television and the smog and all the niggers movin' in around here."

Bradovich stared at him hard.

"Well, they gotta right to live," the dwarf said magnanimously, "but why around here? Knocking real estate values down." He kissed his white plump mouse, drank his lemonade, closed his eyes and patted Bette on her fat behind, a beatific smile on his thin face under the dehydrated banana of a nose. A man in his heaven.

Each to his own heaven, as each to his own hell,

Bradovich thought as he stood and thanked Ben for his hospitality. But Bette was snuggling in Ben's hair now, preparing to bed down to avoid having an hysterical scene, no doubt, and Ben's eyes were still closed in happy, possessive and possessed repose. He was in his very own house, with his very own love and his very own memories of past glories. The fat lady of the circus with her rouged nipples and black twat smiled sweetly with Kewpie doll lips. The Duke played sheriff and bad guy with him. Bradovich no longer existed for him.

As Wizard bent double to pass under the doorframe, Ben growled, "Hey, Golo said to tell you you only got a couple days, then the stay's off. You better get home. Sorry, kiddo, that's the way the screw turns."

Bradovich bumped his head hard, then turned to stare at Ben, but the dwarf had his eyes closed, looked like he was fast asleep. As Wizard slumped dejectedly down the street, two tall square-shouldered operatives in head to toe black shadowed him.

*F*or his family Bradovich fixed a happy face.

Pretended an energy that had left him with the swiftness of an escaping snake at Ben's parting words. He embraced everyone with a fervor that made them stare at him.

"You'll come again soon, won't you, Pop?" Laura asked, staring into his hooded eyes with a concerned look. They'd had only one quarrel, and that had been about the way she wore her hair.

With a teasing smile, he now said, "If you wear your hair pulled back to reveal that beautiful strong face and perfectly shaped head." She shrugged, refusing to bite, and he continued, "Sure, of course, I'll be back. I'm just so elated to be here at last, forgive my excessive exuberance." He tried to relax, to cool it.

The last few hours he spent playing with his grandchildren. He wanted to impress them on his very soul, to impress his on them. At the airport they all embraced and kissed (he almost felt, morbidly, as if it were for the last time), stepped on each other's words as they expressed their happiness at having seen each other after so long an interval. Martin, the calm one, shed a tear; Laura, the excitable one, smiled; Pegeen gave him a last lavender-fragrant hug; Gonzalez shook his hand hard, the pipe jiggling in his mouth.

Bradovich smiled brightly. "Hey, stop the gloom. I'll see you all again soon. Be well, be well all of you." Alexander and Carlotta were given one last crushing embrace, and then he moved on to the ramp.

They waved sadly. He blew them one last kiss as he turned his bulk toward the waiting purser.

On the plane, to shield himself from The Authority's two operatives who sat three rows in front of him, he ducked his head behind an outspread newspaper to conceal his tears. The flight attendants kept interrupting in their customary way, bringing food and drink of one sort or another in relays so that finally he conceded to them, ate and drank his sadness, his bitterness away.

The Authority was hardly to be trusted, was flighty, it seemed, frivolous, gave with one hand, took away with the other. Played games. Made a patsy out of you. A stay was not a reprieve, it was only a pause. The length of the pause arbitrary. Golo had cautioned him. The battle was resumed.

Well, he'd fought them to a standstill before, had had an opportunity to see his children, their families, there was nothing more he could do but renew his stand.

*T*o escape the operatives of
The Authority, Bradovich tried everything. He changed
course in mid-block. Ducked through alleys between stores
or apartment buildings. Hid in doorways. Spanned roof to
roof. Leaped into slowly moving cabs, leaped out again at
random. Nothing helped. They kept pace with him. Thought
ahead of him. It was his territory, but it was also theirs. In
fact, they acted as if they owned it, had owned it forever.

Still, the operatives this time around didn't lay a hand
on him, and for that he was grateful. Grateful? Shit, for what?
To be grateful was yet another concession to them. Besides,
he was aware that they could cold-cock him anytime they
wished.

He refused to concede. They were his enemies. He
would fight them to the end.

He stayed away from his apartment for most hours of
the day. Always ate out, one coffee shop or another. Now was
not the time for gourmet food. Sweeping, dusting, changing

bed linen became chores he neglected. He didn't shower, shave, wore the same clothes—even underwear and socks—every day. He began to stink. Grew a beard. Gray, with a tinge of red. His mother had been a redhead. A Jewish girl from Cicero, Illinois. Every Italian stud in town chased her. Everybody knew Jewish girls were the hottest fucks around.

Beatrice he did everything to avoid. She was a red-head, too, and in late middle age never seemed to get enough. He missed her, yearned for her, daydreamed of yielding himself to her as she had yielded herself so completely to him. He didn't want her pity, her sympathy. This was his own battle, he would fight it out himself. Obstinate son-of-a-bitch! Resting his head on her ample bosom would give him surcease, rest, comfort. To hear her laugh, to listen to her just plain common sense. No, he couldn't, wouldn't abide it.

She, of course, phoned several times every day. She wasn't proud, just dignified in a jokey sort of self-derogatory way to show that life wasn't all that serious. So what? Wizard put her off with a variety of excuses, none of which made any sense. Not a masochist, and not stupid either, she caught on, or thought she did, had had enough, and stopped phoning, but not before exploding. "You're a no good bastard. You think you're too fucking good, too high and mighty, just too grand to join the rest of us, the all-American prick. Go ahead, go ahead, do it on your own!"

He told himself he was avoiding her for her sake—Wizard Bradovich, saint. But he knew it was for himself. He was being hemmed in by the stonefaced shadows of The Authority, and he sought a privacy within himself that he never knew existed.

Then he began to smell the rot, the very stink of himself, and he felt a sudden shame: they were beating him down without a real fight, he had ceded the territory of himself to them. They were beginning to own him. For no reason—perhaps it was the birth of a crisp sunny spring day after many nasty cold gray days—he had a sudden surge of courage, enough to make an abrupt change, enough to reassume his lifelong fastidiousness. Even more so.

He was transmogrified. He shaved his unkempt beard. Had his hair cut and shampooed in a unisex hair salon. Appeared on the boulevard in the garb of a debonair gentleman. A suave giant. Began to wear a homburg, though he hadn't worn a hat of any kind since he was a boy. He bought a cane with a heavy gold crown—gilded iron really. Soon it became polished smooth. The weight of it, the solid tap of it on the pavement flowed upward through his arm and into his heart. It reassured him. People had always stared at him because of his height, his depth of chest and breadth of shoulder, now they saw him all dolled up. When Floriana ran into him in the street outside their building, she said, "You look like you're either going to a wedding or a funeral, Wizard."

He smiled, tipped his hat, and walked on, the cane swung with insouciance. He could not hide from The Authority's operatives, of course. They were on to him. To them he was the same old Bradovich.

Now he reapproached Beatrice but she rejected his attempts at a renewal of their affair, even their friendship. It wasn't because of his previous rejection of her, she was not a vindictive person. She spoke to him about it when he encountered her outside the supermarket.

"Everyone knows about your trouble," she said. "If I could be helpful, I would. I really like you very much. Once I hoped you might ask me to marry you—but not now, it would be just plain nuts on my part. I miss you and there are nights when I wish you were with me, talking to me about your work, laughing at my silly jokes, on top of me, banging the hell out of me with your big thing." Beatrice had an old-fashioned inhibition about saying penis, vagina, clitoris and testicles. "But what's the sense?" she asked forlornly, echoing Bradovich's own earlier thoughts, then smiled sadly.

"The whole business is ridiculous," Wizard said. "The dirty bastards have had me under surveillance now for months and they haven't been able to make a case. Slow as they are, it'll have to be over soon. I don't like losing, never have, and besides I'm clean as a whistle."

Avoiding his eyes, Beatrice said with a sigh, "I hope so, Wiz."

She had given up all hope for him, he could see. Perhaps for herself as well. Her husband of some twenty-five years had left her for an eighteen-year-old secretary in his law office, and her two sons were off to graduate school in the midwest, so she was alone. Refused alimony from her former husband, told him to go take a flying fuck for himself—she had no inhibitions about using four-letter obscenities. She made a decent living as a freelance editor for publishers around town. She specialized in science.

Bradovich followed her with his eyes as she entered the supermarket, the red cloud peppered with silver floating above her head. Soon there would be a cloudburst. Not only was she smart and independent, she had a big sweet ass blooming above her long stems, an ass on which he would have loved to pillow his head again.

Good days and bad. He strolled about the city in new modish clothes from the Tall Men's haberdashery, gold cane in hand, one or another homburg on his head. Went to museums, flirted with middle-aged women looking for a man to replace the one who had died or deserted, saw old movies he'd seen in his youth, ate in fine restaurants. Bradovich had time on his hands. His studio and work he had erased from his mind. Out there somewhere his marble heads stared blankly, his wood heads cried, "Oh, my God!" When one was sold, his agent collected his percentage, deposited the rest in Bradovich's account. Money meant nothing to Bradovich—he had more than he needed. Bradovich hated to go home. Began to understand those un-sheltered derelicts who refused help of any sort from city officials. A home could be a prison, one's fear, a ball and chain. His new clothes were a disguise—the only person he fooled was himself. In truth, he was the only person he was trying to fool.

The Authority's operatives continued, for whatever reason, to be benign, and Bradovich began again to ignore them, to become his old self. He discarded the gold-headed cane, the homburg. Instead of Savile Row-designed suits, he returned to his sculptor's uniform: an old pair of corduroys, a faded blue work shirt, a patched sweater. Perhaps he'd go downtown to his studio, see that everything was in place. Run his hand over a block of granite. Caress a slab of Norwegian mahogany, a stick of Gabon ebony. His heart moved, a tear appeared in each eye.

He entered the subway several blocks south of his apartment building. Hadn't been in that hell in weeks. What a stench! The dust made him sneeze. Not too crowded. A downtown train had just left. Leaned against a steel vertical post. A couple of eighteen-year-olds in patched blue jeans, thermal jackets, Yankee baseball caps, unlaced dirty white sneakers, slunk past him, swivelled heads to give him the eye, too big to hassle, continued on. A train roared in on the uptown side. A boy and girl leaned against the opposite side of the same post as Bradovich, girl with her back against it, boy against girl. She was an Asian, squat, flat face, bow-legged, the boy Nordic, tall, slim. He leaned against her hard. The inscrutable Oriental laughed, showing large white teeth. "You'll shoot your load," she said in perfect English.

"Wait till we get to my place." The boy laughed in perfect Swedish. Leaned harder into her, began to rotate his skinny behind.

"If you can't hold it, kid," Bradovich said, "there's a bench down further. It's also darker. There you can screw to your heart's content." Wizard's tone was quiet, but his largeness scared them, and they left without a word, leaned against the grimy tile wall, talked now, perhaps abashed.

A middle-aged couple, fat, sweaty, spoke to each other in Spanish. Both had overflowing bags with string handles hanging from either hand. They had been out food shopping. The man's belly hung over his belt. She wore an old house dress under an open gray woolen coat and cardboard shoes. Sweat beaded her flat forehead, her tired eyes blinked drops away, and she was telling her husband to shut up, he had complained enough. "I can't do everything myself. You pump me full of children, I work like a donkey day and night." The man turned his back to her.

Another train on the uptown side. Bradovich began to talk to himself. "Things don't get better, they get worse. Where the hell is the stupid train?" The Spanish-speaking couple stared at him. He smiled and bowed his head to the lady. She smiled back. The Spanish gentleman muttered under his breath, something about a *pinga grande*, and Bradovich smiled and bowed to him as well. Because he was a big man, everyone thought he had a big prick. They didn't seem to be aware that every big prick had a little prick hiding within it.

Two heavy-shouldered gorillas joined the growing crowd on the downtown platform. Both dressed in blue serge, flowered ties, pearl gray fedoras, black pointed shoes. They stopped to ask the Hispanic couple if the train stopped at the railroad station. *"Si, si,"* the woman said. *"Terminal grande.* Shuttle." The gorillas didn't understand. Bradovich

161

explained it to them, that there were two railroad stations, and how to reach both. "If the train ever comes in." The men nodded, moved back against the dirty tile wall. One of them began to sneeze from the dust. He sneezed seven times. A drunk reeled by, singing to himself. Found a wall, sat down, his feet stretched before him, a wide grin on his face. A happy drunk. Resumed singing, "Tea for two, and baby makes three, you and me, let's go live in a tree." Then laughed uproariously. The two young punks in thermal jackets applauded. Bradovich stared at him. Enjoy yourself, buddy, while you can. Soon you'll be eaten up alive from the inside out.

The platform livened up with young girlish voices and rough boyish laughter. A school day had come to an end. Bradovich looked at the young girls, first surreptitiously, then frankly ogled them. They were high schoolers, most of them wearing jersey leotards and micro-miniskirts wrapped around their waists, their young breasts showing tiny nipples under the tight jerseys. One of the girls, a bright, brash kid, asked what the shit he was looking at, and he turned away, embarrassed. Damn, how he loved them. He could eat the lot of them.

He began to pace between two posts, anxious for the train to arrive. He would soon forget what he was waiting for. To get to see his studio. To smell it. To touch a piece of stone. Run his palm over a stick of wood. Another train squealed in on the uptown side. He and others on the downtown platform began to lean over the edge to see if their train was coming. The uptown train departed from the station. As it did, a shout went up, the lights of the downtown train appeared coming around a bend, and the crowd on the platform moved en masse toward the edge. "At last!" Bradovich said aloud.

The train's steel wheels began to whistle as the engineer slowly applied the brakes. As the train approached, the two gorillas who had been standing mutely against the tile wall came alive with a spurt of energy, pushing boys and girls aside as they neared the platform's edge. Bradovich turned to admonish them to stop their goddamned shoving. They were coming so fast they couldn't stop their onrush and both hit Bradovich with their burly shoulders and he tumbled onto the tracks, the train rushing down on him. Those on the platform heard him shout "Oh, shit!" before he was lost from view as the train hurtled past and then came to a stop. There was a great deal of screaming and yelling as the conductor stepped out, soon followed by the engineer. When the Spanish man turned to point out the culprits to the policeman who appeared, they were not there. There was only one thing to do, since more than half the train had passed over the spot where Bradovich had fallen, to send the train ahead. Everyone boarded the train, the policeman and now his partner remaining, and the train moved on to its next destination.

When the train left, Bradovich's body was not visible anywhere on the tracks. One of the policemen jumped down and looked around. He heard a groan. He found Bradovich's huge body pressed lengthwise under the platform's overhanging lip. Bradovich was facing forward. His sweater had been ripped clean down the center its full length, and the tips of his ankle-high walking shoes were torn open. He appeared to be physically unhurt except for several minor bruises on his hands which had supported him when he hit the tracks. The cops helped him onto the station platform. His corduroy trousers were dusty and torn. His face was ashen. His eyes stared. The cops got him to lie down flat on his back, and held his feet up so the blood could run back to his head. One of them removed his blue jacket and covered Bradovich with it to keep him warm in case of shock. An ambulance had been called. When the medics arrived, they found Wizard's pulse normal.

After giving the policmen a full description of what had occurred and of the two gorillas—he spoke very slowly—he refused any further help, but sat up, then stood by himself, thanked the two cops—young, fresh-faced, in the spring of their lives—and the medics, and walked slowly out of the station, refusing help of any sort. He looked dazed as he trudged back toward his apartment building on the boulevard. His eyes stared straight ahead, his shoulders sagged, he could barely lift his feet off the ground, he just sort of shuffled. He pushed by Slim who saw that the man was in pain, found his way to the elevator, pressed the button for the eleventh floor. At Beatrice's door—he wasn't called Wizard for nothing—he rang the bell. He heard her footsteps, he was sure she was eying him through the aperture. He could hear her sigh. The chain and lock rattled, and the door opened.

"What the—?"

Bradovich simply fell to his knees, and then to all fours and crawled past her into the hallway. There he passed out.

*B*radovich was too heavy for Beatrice to move. She phoned Slim downstairs. Together they dragged Bradovich to her bed and together managed to raise him, first head and back, then thick legs. Bradovich opened his eyes, but said nothing.

Beatrice thanked Slim. He left, shaking his head, panting heavily.

Beatrice undressed Wizard. Washed him down with warm water and soap. "What happened?" she asked him.

He told her.

She cursed them, it, whatever, nursed him, cooed over him, cooled his brow, held him in her arms, his head couched on her deep breasts. "You're a brick," he said.

"And you," she said, "are a fool."

"I know," he said, snuggling closer, "but I'm getting smarter by the minute."

He fell asleep.

In Bradovich's apartment, where Beatrice had gone to fetch fresh clothes for him, she encountered Mrs. Kastner at her window. First they made faces at one another, competing, and Mrs. Kastner won with ease. Then they spoke.

"What do you think?" asked Mrs. Kastner.

"I don't know what to think," replied Beatrice Holden.

"They almost got him this time, didn't they?"

"Another half inch and he'd have been split open like a watermelon."

"The next time they'll get him, you'll see," Mrs. Kastner said with big city finality.

"Don't bet on it. He's a tough bastard."

"You left out a word. A tough *old* bastard."

"Not so old," Beatrice Holden said, then smiled with a slight turn of her lips.

"Is he still good?"

"Better."

"He's really built."

"Who knows better than you, my dear Mrs. Kastner."

The latter stuck her thumbs in her ears, wiggled her fingers, showed her tongue.

Beatrice laughed, turned to leave.

Mrs. Kastner sang, "It's only you, there's no one else. . . ."

Beatrice fed Bradovich, cosseted him, scolded him, made love with him, acted like a wife. And he like one of those husbands who rarely talked.

When she thought he was fully recuperated, she said, "Call Bennett Pollack, the famous criminal lawyer. He'll help you. He's never lost a case."

"Golo's still on it."

"Another mouthpiece won't hurt. It's good to double-check."

"It'll just be a waste of time."

"Yes, I know, there's so little time to waste," she said. "I know the man, once had an affair with him after meeting him at a bar association dinner I attended with my ex-bastard. Incidentally, I believe all men are bastards, which of course includes you. If you don't call Pollack, I will."

"Okay. But don't you dare go near him."

She smiled. Mona Lisa herself.

Bradovich felt as if it were an admission of guilt merely to dial the famous lawyer's phone number.

"There has to be good cause shown for surveillance," Pollack said. "And no one can bug you without a court order. Assault and battery is a felony. Attempted murder is a capital crime."

"They even have a key to my door. They interrupt television programs to give me warnings, threats. They never stop shadowing me. They've tried to kill me three times. They entered my studio and hung one of my wood heads in a noose. What the hell can I do about it?"

"Oh," Bennett Pollack said, "*that* authority! There's nothing I can do for you. It's between you and them. The most powerful influence peddler in the city is Golo the Gimp. If you can get—"

"He's my friend. Has been trying."

"If he can't, no one can. Good luck, Mr. Bradovich. No charge. Best to Bea Holden." He hung up.

Bradovich shrugged. The result had been no more than he'd expected.

Sad, dejected, Beatrice and Bradovich embraced. She cried. He was tearless.

He hugged her tightly, crushed her, raised his old head to the heavens, said, "What can I complain about?"

Beatrice Holden kissed him sweetly on the lips.

*I*t was Sunday afternoon. Bradovich was dolled up, even had his cane under elbow. Beatrice had to stay behind, she had a deadline to meet with a manuscript. She was serious about her work. Wizard tried hard not to think about his.

He was to join his old friend Strayhorn at the old renovated concert hall. Madeleine Dearing was going to sing Brahms and Schubert *lieder.* It was a glorious day, the sun high, the breeze gentle. The Authority's operatives had already removed their gray topcoats, appeared only in their black suits and black fedoras.

Bradovich decided to walk via the park. It was only some twenty blocks, about a mile. The park was crowded with promenaders, children on skates and bicycles. There were joggers, speed walkers, and just plain lollers and loiterers.

As Bradovich passed through the entrance, he saw a woman sitting on a bench all alone. Beside her rested a heap of packages. She was stylishly dressed, appeared to be wealthy. Several glittering rings adorned the fingers of her gloved left hand. Her brown gray hair cut close to her well-shaped head rippled in the breeze. Sharp-nosed, flat-cheeked.

As he passed in front of her, he recognized her, hoped she didn't see him, but saw she was crying. He continued on briskly, the cane held close to his body. After striding on for some fifty feet, he sighed, hesitated, spun on his heel and retraced his steps. "Poor kid."

When he stood before the weeping woman, he asked her if he might be of any help. She raised her head and tearfully shook it, tried to smile, but when he stirred as if to leave she began to weep more violently. He turned back to her and, first placing the packages at her feet, sat down beside her.

He took her hand in his and stroked it gently, but said nothing. They sat thusly, the giant of a man and the weeping woman, for several minutes, her sobs violently shaking her, he holding her hand and stroking it. Seeing that she wasn't letting up, he moved closer to her and placed his arm about her quivering shoulders and pulled her gently close to him until she could rest her head on his chest. He still said nothing, merely held her in this way.

After a few minutes, the passion of her sobs began to subside, she didn't move away from him, just sat there, feeling the warmth of his body, her head on his chest, his arm about her.

From his jacket breast pocket Bradovich withdrew a large handkerchief and gave it to her. She raised it to her lips, patted softly, wiped her face, never once looking up at him. There were greater intervals between her sobs now, the waves of the sea subsiding as the storm waned. The wind passed on, the sea rippled but slightly.

170

Then a sibilant sigh, followed by absolute calm.

They sat for many minutes, her head resting on his chest, his arm about her, one huge Bradovichian hand holding her slender one.

Finally, he said, "Why, Francine?"

"Because," she replied.

"Of course," Bradovich said.

She rose, shook his hand, picked up her packages, smiled quietly without revealing her teeth—she'd always had bad teeth, he wondered if she still had them—for the first time peered directly into his eyes, nodded her head, said quietly, "I should never have left you, Wizard. It was a bad mistake."

He shrugged his shoulders.

"I hope you overcome your troubles," she said—so she knew, too—and continued into the park.

Bradovich ran out to grab a taxi, he was late.

The concert hall was filled to overflowing. Even the standing room was sold out. Madeleine Dearing was one of the great sopranos of the world. Many years before, she'd become mute from a form of depression brought on by a personal grief about which Bradovich was ignorant. She'd become mute and terribly obese. Wizard's friend David Strayhorn, himself at the time a lean wolf, had encountered her and become obsessed. His obsession had been transformed to love by Madeleine's inner beauty. They married, had children. Strayhorn, now sitting alongside Bradovich in the concert hall, carried a smug paunch, a balding head and a proud smile as Madeleine held the stage, filling the hall with Schubertian and Brahmsian beauty. When she sang Brahms' *Cradle Song*, she moved Bradovich so profoundly he decided then and there to ask Beatrice to marry him and to relocate on the West Coast so that he could live near his grandchildren. He was certain she would say yes.

After going backstage with Strayhorn to congratulate Madeleine, Wizard decided again to walk home through the park. It was a perfect evening, after a perfect afternoon.

And the stroll through the park was perfect. No rowdiness, no confrontations. Couples arm in arm spoke quietly, children laughed, young men and women were polite when they were in a rush to pass you.

Still, no day is perfect. As Bradovich was about to enter his building to pick up Beatrice for a dinner date, two stonefaces barred his path. He raised his cane, "I'll beat your fucking brains out, out of my way."

"Your trial will be coming up in three days. Be prepared."

"At long last, you bastards," he shouted as they turned their backs on him.

*T*hey met, Bradovich, Beatrice, Floriana and Golo the Gimp in Floriana's lush living room with its deep sofas, divans, poufs, thick carpets, and elegant velvet drapes.

Golo was stretched out his full four and a half feet on the curved sofa, his large head couched on Flor's luxuriant lap. She was dressed in a carmine satin pants suit so that under the soft lamplight whenever she moved every roundness and secret niche of her body was glossily revealed. Floriana and Golo were to be married in a week at a small private wedding. Beatrice and Wizard were to stand for them. Floriana had already retired and given her customers' list to Consuela as a gift. Golo was creating a new magic act, and Floriana was to be his stage assistant. The Gimp had been considered one of the premier magicians in the world years ago, but he had given up the profession to become a mime.

Beatrice, dressed in a denim skirt and yellow sweater, and Bradovich in his worn corduroys and faded blue work shirt were sitting close together, filling every inch of a sturdy Victorian love seat. They were holding hands.

It was the evening before Bradovich was to meet his accusers before The Authority, and they were discussing strategy and tactics.

"Lying isn't so terrible," Golo was saying. "On occasion it just has to be done."

"I don't even know who the hell I'm to lie to, or what I'm to lie about," Bradovich said quietly, bringing Beatrice's hand to his lips and kissing it. Ever since he had informed her of his coming trial, she'd been nervous, silent, reserved. Worried sick.

"We are what we do," Golo said, pressing his head deeper into Floriana's lap. She adjusted her round thighs slightly to make it easier for him. Bradovich had a vision of him turning his head and lapping at her like a docile dog. "I'm a mime, a prestidigitator, you carve a piece of marble and pretend it's a head. We lie all the time."

"Don't confuse art with life, for Christ's sake. You know better."

"If you were to marry Beatrice," Floriana said huskily, "you might think differently."

"Then I'd be living under a perpetual red cloud," Bradovich said, then laughed, as did Floriana and her lover. Beatrice didn't laugh. She didn't believe that the cloud under which she lived day and night was a laughing matter. It was her crowning glory and part of her uniqueness.

"I don't see what the hell you are all laughing about," she said with not a little rancor. "I finally found someone to spend time with, to live my life with, and now what?"

"I'm sorry, Beatrice," Wizard said, giving her a squeeze. "I need something to laugh about. Anyway, I love your hair."

"Liar," she said, gently squeezing his thing as she

called it in the semi-darkness. "You've been after me to cut it, to shape it so that my gorgeous skull can be revealed. But why talk about my hair when there are more important things to talk about—like the black cloud under which you live day and night?" Beatrice was if nothing else a realist.

"I've had it up to here," Bradovich said, bringing the edge of his large hand to his throat. "Sometimes they're so close, their shadows fall in front of me and *I'm* shadowing *them*. Then who is shadowing whom I want to ask, like who's the victim and who the victor."

No one answered, of course.

"Lying," Golo returned to the subject, "has saved many a man or woman's life. When life is arbitrary, one's morality has to become arbitrary. Otherwise this species will die out. Wouldn't you have lied to the KGB or the Gestapo if taken prisoner?"

"If necessary, of course."

"There you have it—if necessary," Floriana said.

"A lie's a lie like a man's a man," Bradovich said in a loud voice. Tired of it, tired of the whole bloody thing. Beatrice patted him gently on the thigh. "Should I plead guilty if I'm innocent?"

"No one's innocent," Floriana whispered.

"If necessary," Golo said loudly.

"Why, goddammit?"

"Why? Because," said Golo. "Why because? Therefore."

"I want you to live," Beatrice said softly. "What's a little lie if you can buy your way out with it?"

She slunk low into the corner of the love seat, her head tilted upward, her eyes slitted. Bradovich could see the tip of her not short nose. Was she pretty or wasn't she? He could never decide. But when they made love, when her face was flushed and swollen with passion, she was exceptionally

175

beautiful. He was getting to love this woman more and more. He, too, needed someone to spend time with, and not just at night, and especially not just in bed. He had to admit he was getting old. Getting? What a laugh. Already there. The Authority was beating him down. Everyone was beating him down. Beatrice, too.

"Wherefore, whereas, therefore and hereinafter. Why pettifog about it?" Golo sighed.

Bradovich peered intently at the Gimp. His eyes were closed, a contented little smile scribed his strong lips, his head sunk deep into Floriana's red-sheathed crotch. If Golo prayed, it was at the altar of cunt. Why was he so intent on Bradovich's pleading guilty to a charge so far not yet even alleged? Why, indeed? Bradovich wondered. Had the Gimp ever been in the same situation and come through with a plea of guilty to some unspoken charge? Or had he, in order to overcome the burden of his physical uniqueness, felt it necessary to corrupt his life with lies and bribes to curry favor with the mighty and saw no earthly reason why his friend Bradovich should not do likewise? Golo's fame had given birth to great power which the Gimp had learned to use wisely, but also ruthlessly, and with his power had come corruption. When you're corrupt, Wizard thought, there's that terrible need to entice everyone else into corruption. Why should the innocent stand alone, unique, above it all? Corruption loves company. Or perhaps it's just that Golo's being a damned good friend.

"We haven't moved forward an inch," Beatrice said, patting Wizard's huge fist which covered her thigh.

Is everything you do done for selfish reasons, however secret or innocent? Is Beatrice thinking of me or is it her own needs she's trying to protect? Under every stone there's a lizard.

"I'm holding the fort," Bradovich said.

"Under my cloud," Beatrice said, laughing, joined by the rest of them as they rose to make their exits.

Beatrice and Wizard to her apartment—bugless and shadow-proof by The Authority's arbitrary rules—and Golo and Floriana to the dressing room set up and furnished like one backstage because Golo felt happiest there, the great star anticipating the immense pleasure that comes with bringing down the house. Break a leg, Golo. There on the casting couch Floriana didn't have to pretend to passion because Golo was an artful and ardent lover—and, even more importantly, because she loved him, had relegated her impure love for her brother to the distant past.

In Beatrice's apartment, Bradovich found no peace. She railed at him nonstop (he had to admit he enjoyed her nagging, found it proof of her love), importuning him to concede to Golo's logic, to bow his head for once, to admit guilt even though The Authority had not turned up a scintilla of evidence.

"No, I'm sorry, Beatrice. Arbitrary authority requires constant feeding, and I won't give them a turnip. Not even a turd. That's the only way to beat them."

She continued on as though he'd said nothing. Ox, stubborn mule, dumb bastard, self-destructive, suicidal, idealistic, romantic, how stupid can you get?

By three in the morning, he became too fatigued to argue further with her, so he seduced her into bed. At the height of her passion, Beatrice cried out her love for him, "Wizard, you son-of-a-bitch, you had better!"

Yes, he had better.

*D*ressed in his work clothes, seated in the dark near his early head of Valerian, his hand resting on its cross-hatched surface, Bradovich awaited the arrival of The Authority's operatives.

The appointed hour came and went, and they did not come to take him.

At first he was impatient, kept throwing obscenities at them, but as the minutes and then the hours passed, a langour overcame him, a peaceful lassitude, and he fell asleep.

It was the morning sun which came to wake him. He felt rested. He laughed. The bastards are afraid of me. His refusal to bend had defeated them.

Wizard stood, adjusted his clothes, washed, combed his hair, and went to Beatrice.

"Bradovich! Bradovich!" she shrieked with delight, with happiness.

Downstairs, in the brash sunlight, two stonefaces leaned against the building brick, lit cigarettes, and impassively puffed away.

*F*loriana and Golo the Gimp, she in gold lamé, he in actor's electric blue, were married in Intercontinental's pink-carpeted and pink-walled small ball-room on the eightieth floor. The city was lit by a dying red-gold sun. At the point where light and shadow met, the sun transmuted the pink to a blood red.

Beatrice wore an off-the-shoulder turquoise dress to show off her red hair no longer a cloud but brushed back and cut close to her oval head in a French coup, her nose proud, her forehead pure, so that Bradovich in afternoon gray and black, austere, gold head of cane gleaming, was very pleased. Also in attendance were Consuela in cream white and her present lover, a contortionist friend of Golo's, a wiry hiero-glyphic figure of a man also in cream white—Bradovich noted it made him look like a white-enameled Giacometti—and Mr. Sam Rabinowitz, Golo's agent, who turned out to be a young Jimmy Stewart, and his new bride, his third, an up-

and-coming modern dancer already touted as the new Martha Graham. Bradovich could not stop ogling her even when Beatrice pulled energetically on his sleeve. Sam and his bride were dressed in distressed blue denim suits. Madeleine Dearing sang several Handel arias. Strayhorn stood beaming at the side. Judge Constantino Romero Bellini presided at the nuptials.

Below, at the entrance to the lobby of the concrete and glass monolith, as if guarding the door, stood two of Bradovich's shadows, stonefaced, dressed in black as usual.

Mr. Rabinowitz provided smoked Scotch salmon, Russian caviar, bagels, cornbread, and pumpernickel, as well as the finest champagne.

There were toasts all around. While Golo regaled the other guests with dirty jokes, Floriana spoke to Bradovich on the side.

She had tears in her eyes. "The night they were supposed to come to you, they came to me instead."

"What for?" Bradovich asked, cheeks paling. "What the hell did they want?"

"They said I was in trouble, was under surveillance, on my own recognizance, not to leave the city."

"The bastards! The dirty bastards!"

"One of them was the man who had sodomized me and ransacked my apartment."

"What did you say?"

"I told them to go to hell, then called Golo. I don't know what he did, but he got a stay so we could get married and go on a honeymoon."

"You have to fight them all the way, Floriana, that's the only way to beat them."

"I know. But damn, just damn!" She was fighting to hold back the tears, her beautiful face taut with pain. "I finally made it, got out of *the life*, was able to give up my brother, found a man who respects me and loves me, and now this."

All the time they were talking, Golo, who was throwing out quips without a stop, kept his eye on her. Finally, he broke away, and came to take her. "We have to go now, honey," he said, taking Floriana by her hand, bending his large head backwards so he could look up into her tear-filled eyes. "Don't let them see you crying. Give them a smile. The show must go on."

Floriana wiped her tears away with a small handkerchief, forced a smile. She was going to make a great trouper. In an aside to Bradovich, Golo said bitterly, "If I have to choose, I will, of course, choose Floriana. You are on your own now, Bradovich." With that he took Floriana by the hand, waved to everyone, pulled her to the door and they left under a barrage of rice and *bon voyages* from the guests.

Bradovich just stood flatflooted, staring into space. What did Golo mean by that? Was he blaming him for Floriana's trouble? Was there a choice?

Back in Wizard's apartment, where he was gathering some clothes to take to Beatrice's flat, she suggested that they also get married. She grinned after making the suggestion, her pointy nose rounding at the tip. To her astonishment, Bradovich, still under the magic spell that weddings have on romantics, said, "Sure, honey, why not?"

Whereupon, Beatrice slid a cassette into the player, and to a raucous disco tune began to do a strip tease as Bradovich watched open-mouthed.

As she began to remove her bra, and Bradovich felt the bird rear its head behind his fly, the tune was interrupted by a metallic voice which said, "Mr. B., tonight at 2. Your time has come."

*B*radovich was ready for them when they arrived. It was two in the morning, the city asleep. He was wearing his work clothes: corduroys, denim shirt, and the heavy work shoes with their cupped steel toes that protected him should a granite block slip and fall.

"Just don't lay a hand on me," he said with menace in his voice. "I go willingly—want to get this fucking matter over with."

They said nothing—two stonefaces in black *cap-à-pie* except for starched white shirts, button-down.

The city was dead quiet except for the occasional squeal of an errant taxi tire or the scream of a vigilant observer at the dying of a loved one. Bradovich's heels echoed in the city's concrete canyons. Those of The Authority's operatives were sibilantly muted.

They led him into the park, took a diagonal course toward the east side of town. The trees whispered. Rodents skittered across their path from one bush to another. Snakes uncoiled from under rocks. Birds slept. The moon was a squiggle of white ink. Somewhere someone cried out. In pain? In ecstasy? A clock in a tower struck two-thirty. It is at two that most are taken. Bradovich laughed aloud. "Well, I've gotten past that hurdle, haven't I?"

The two stonefaces shrugged noncommittally. Bradovich had turned out to be one of their most obstinate cases. Yes, he was a gnarled oak of a man, a muscle, a tight knuckle-scarred fist. He took what was given, returned in equal portion. "I believe in the laws of physics," he said. "If you fuck me, I fuck you. Life is not a football game for amateurs. It's for real. If you want to play for fun, stay in kindergarten. Be a toddler for life. Doodle on a pad at the telephone and pretend you've made a work of art. Play kid games, go ahead, you fucking empty head you. The earth's whirling like a madhouse, people are dying by the millions, and you want to play amateur games. That's for kids, you fucking idiots. Grow up. It's a serious game. You play it to win, otherwise you go down with the dinosaurs. You're standing at an open window and looking down. You have a stone in your belly and you don't even know it. You bend out the window and stare down at the concrete. The stone in your belly rolls toward your head and its weight keeps pushing you out the window and down to that concrete. You yell, Step back, you fool! Stay cool. Avoid panic. Your body remains static, immobile. You look down and see the tiny men and women scurrying about. You lean over as far as you can without toppling down into that concrete, that stone in your belly now choking your gullet. So you're standing at the edge, it has come to that at last. Has it all been worth it? you wonder. Have you played the game as it should be played? Have you formed that block of marble and made it say what you have always wanted to say? Or is it choking you to death? Those blank faces out there, staring, mute, carrying their pain in their frozen hearts. Stone. Polished stone. What are you hiding from? Why are your eyes blank? Why do I hide from the world the way a child hides under the blanket in his fear of the dark, becomes more frightened and presses his eyelids tight, tighter, until he sees the falsely induced suns you see when you press your eyelids hard together. It would be simple, it seems, to escape the burying blanket and dare the dark for a few moments and stand at the window to observe the real stars in their flight, or, simpler, less poetic, to dare the few dark, frightening steps to the light switch on the wall and expose yourself to man-discovered light and then to smile at one's own silly, embarrassed smirk in the mirror. The mirror image would perhaps reveal a frailty, a vulnerable tenderness. The cool gentle hand of Valerian in the middle of the night to calm your tormented sleep. No, you say, taking one step closer to the edge. There is now no more to go. You have not dared face the risk of living for so long that now you stare into the deep, howling, very hard face of death. You stand at the edge, the moon a glow worm, and stare down to the concrete and become frantic, and becoming frantic you can still wonder at the revealing thought that life and death are siblings, twins, very real kin, all else superstition. One more step to nothingness, that's all. Go now. You stand at the very edge and do not feel the wind or even hear its howl. *It is up to you alone.* And you say, No! No! and you step back from the edge."

With a quick movement, Bradovich spread his long, muscular arms and clasped each operative's stonehead in his massive fists. It would be easy for him to smash their skulls together, smash them like cantaloupes, but he would still have to face The Authority's judges. He unclasped them, went quietly and swiftly. The great clock in the Grand Old Terminal showed 3:05 when they entered.

First to an overflowing baggage room smelling of old

185

leather and canvas with a makeshift desk behind which sat the pre-trial magistrate in a black robe, gavel in hand. Another stoneface. A mid-level bureaucrat. With metallic voice, "You are Stephen Wizard Bradovich. Who are you?"

Bradovich stared at the magistrate under the dim light. Blank face.

"I am who I am. If you don't know me, why am I here?"

The gavel rapped authoritatively. "Who are you, Mr. B.?"

"I am Stephen Bradovich. My father was also Stephen Bradovich. My mother was Mary Golda Cohen. I have a sister named Sybil Bradovich. She constructs stained glass windows for cement block churches in which penitents eat the flesh of Jesus H. Christ. I am in my seventh decade. Who the fuck are you?"

"You are making things difficult for yourself, Mr. B. Go on, who are you?" Monotone, metallic.

"I was a professional football player, defensive end. Tough. I am a sculptor of stone, marble, granite heads. I also carve wood heads from East Indian ebony, bubinga, Gabon ebony, and cedar six by sixes. I've done your head a hundred times. My heads sit in the museums of the world. I was married twice. My first wife, a vaudevillian, divorced me to marry a glitzy painter. She has since regretted it. Not I. My second wife, Valerian, a librarian, a guardian of our history, a reader of books, a wise woman, died several years ago. I have two children and two grandchildren. I am now preparing to marry another woman, Beatrice Holden, also a reader of books, an editor. She is a very funny lady who makes me laugh. Golo the Gimp and his wife, Floriana, are my friends. I have but few others. My work is shown throughout the world. I am a free man. I know the difference between history and the smell of skunk."

186

Bradovich smiled, said no more.

The magistrate stared at him impassively. "Now tell me who you are, Mr. B."

"I've said enough, more than enough."

"Golo the Gimp is not as omnipotent as he believes. He cannot help you here."

"I don't need his help," Bradovich yelled, caught himself, continued quietly. "I stand on my own feet. Always have. Always will."

"An empty boast, Mr. B. The nub of it, sir, I want the nub of it."

Bradovich conceded. "A man, a woman, never outlives his, her, childhood. Experience forces adjustment, unless, of course, you're a total fool. Time puts a patina on it, a soft verdigris. But every time you make a new turn, there it is, your silly childhood. When I was a kid going to grade school, I used to join with other Hunkie kids, Croats, Polacks, and the like, to gang up on the little niggers and kikes as they ran through the railroad underpass on their way home. We'd yell, throw stones and garbage at them, laughing all the way. We were brutal, cruel—we were the toughest kids in town. Hey, you little nigger bastard, you wanna eat my shit? What great fun we had. There was one skinny Jewish kid I particularly liked to pick on, trip him, kick him in the ass until he fell on his face, and when he tried to stand, step on his fingers. A big-nosed skinny kike.

"One day, after school, to our amazement they had banded together, the little kikes and jigaboos, built a barricade across the opening to the underpass, stocked up on stones and bags of shit, carried baseball bats, and beat the living hell out of us. That big-nosed skinny kike, he carried a quirt, a horse whip, and he whipped me until I bled.

"I straggled home, my back in shreds, my nose bloodied, with a shiner and a lump on my head as big as the rock

187

that made it. My uncle was there, my father's brother. Where my old man was an oak, my Uncle Walt was a reed, listened to classical music, went to art galleries, even read a book—in our family a real freak. After I finished the news bulletin of the war between us good guys and the army of kikes and jigaboos, my uncle just picked me off the floor and threw me against the wall of the kitchen. 'You ignorant little bastard,' he yelled, 'don't you know yet you're a Jew, too?'

"My father jumped to his feet and was ready to kill his brother, but my mother, a little mouse of a woman once pretty now plain, had stood in the doorway and heard everything, and now she ran crying to her bedroom. So it was then I heard where I came from. My mother had been born a Jew but had been considered dead by her orthodox parents when she married a gentile, and my father's parents had little to do with us. Now I knew why my foreskin was cut on the bias.

"So I ask you, your Honorable Stoneface, who am I?"

The pre-trial magistrate didn't flicker an eye, and Bradovich believed he had just wasted his breath. However, his Honor dismissed him with a nod to the escorting operatives. As Bradovich was led into the great trial hall, he turned to the magistrate, and yelled, "You still don't understand, do you, you dummy?"

The immense and majestic center hall of the Grand Old Terminal with its extremely high ceiling and marble walls was crowded with stonefaces sitting in serried rows of folding chairs. Fedoras not removed, but pulled down squarely over their temples. In the remote corners of the hall, crooks, crannies, and niches were filled with the unsheltered, some dead drunk, others oblivious as they sought sleep. The terminal was closed every night from 2:30 to 5:30 in the morning, but here was this bloody court disturbing them. The day had been long and arduous for them, and they

were damned if they would allow the proceedings to interfere with their rest and sleep.

The three judges—the chief in the center wearing a
vermillion robe, the one to his left a blue robe, and the one to
his right, a woman, an orange robe and long gold pendant
earrings—sat on a banc erected high above the circular
information desk. The judges were arguing among themselves, gesticulating forcefully in their heated debate.

Bradovich stood straight and tall, a heavy-shouldered,
deep-chested man, his teeth clenched, his jaw jutted, the
image of the young Accused a model for his behavior. Sure
enough, there at the side of the banc stood the bailiff, a meat
cleaver on his shoulder. At the side and in front of the judges'
bench, behind a sleek stenotypy machine, the beautiful old
woman who had come slate in hand to ask him why, nodded,
half-asleep. The two stonefaces, twins he now noted under
the bright lights, who had escorted him to his trial had
removed themselves and now stood each at opposing ends of
the raised banc where the three judges continued their argu-
mentation. Suddenly the judges seemed to come to some
agreement, stopped arguing and gesticulating and turned to
read the voluminous papers on the desk before them. Bradovich assumed they were the reports on his surveillance.

The assemblage of stonefaces sat quietly, stonily, and
Bradovich raised his head to stare at the blue painted sky and
blinking stars and constellations above him, the Archer
aiming his arrow at Pisces. They should have used Chagall,
Wizard thought, though he wasn't among the great admirers
of that painter. Tiepolo would have done the job best of all.
Now *there* was a painter, but of course *that* was a time.
Painters and sculptors worked with arrogant fist. Power and
color. Afraid of nothing.

On the east wall of the great terminal, foursquare in

all its natural grace, stood a giant jaguar ready to pounce.

Bradovich examined the great hall unafraid. If there was any doubt, he concealed it from himself. The meat cleaver he ignored. The severed head of the young Accused he blanked out. He saw her whole in the flickering light of the tunnel dignified, unafraid.

He pulled his head far back so he could observe the three judges high above him. Just a trick to make me feel small to their high muckamuckness. Who the fuck do they think they're kidding? It was taking the three judges a long time to read their papers, and even the serried rows of stonefaces became impatient and soon there was a low murmur of voices, like the sea at a time of unruffled calm. In the far corners the unsheltered snored, babbled to themselves, gurgled booze from bottles in brown paper bags.

As the murmur crescendoed to a roar, the chief judge, the vermillion-robed figure in the center of the three, raised his head and observed his audience with cold eyes. The bailiff raised his meat cleaver high above his Mohawk-shaven head and called out in a stentorian voice, "All quiet in the courtroom."

"Sssh, sssh," ran through the hall like a shrill gust of wind. Quiet resumed. The chief judge returned to his papers.

Bradovich realized it had been many months that he'd been under investigation and just the simple daily reports would constitute a mass of documentation. When he woke up, what he wore, what he ate, whom he spoke to, when he went to Floriana's, his meetings with Beatrice, his encounters with Mrs. Kastner, when he jogged, when he walked, when he shat, what he thought, what he didn't think. One could be equally guilty by omission as by commission. With a panel of judges appointed by a power as arbitrary as The Authority there was no telling, since there was no established law to be used as precedence. They fucked whom they pleased, how they pleased, when they pleased. Anything goes was their motto. Justice was not only blind, it was deaf and dumb. It was an oxymoron. What specifically had he done that no one else in the world had not done and would not do and would not continue to do? He excluded the mass murderers of the right/left, the wanton killers, vandals, amoral peddlers of mendacity, the avaricious amassers of gold, jewels, property, the kneelers to power, the power hungry, the asskickers, the brown nosers, the asslickers, the buyers of a little time. He was not among them, at least so he thought. He was his own man. Went his way. You leave me alone, I'll leave you alone. Don't exploit me or I'll kick your teeth in. Give me what's mine and you can keep what's yours. Just don't shit on me, I warn you. Wizard Bradovich.

The judges had again stopped reading their papers and were joined in further argument. The serried rows of stone-faces, their reports being questioned, again became impatient, perhaps even infuriated in their stony way. Began to argue among themselves. The boozers in the corners began to hiss. Bradovich saw the hands of the huge clock approaching 5:30, the terminal opening time. He smiled to himself.

He, who had not been dubbed Wizard for nothing during the football wars, understood well the meaning of hesitation in the ranks of the enemy. Confusion. Now's the time to strike.

"Fuck em!" he shouted so loudly it echoed off the marble walls of the Grand Old Terminal.

Everyone in the courtroom raised his or her head to stare at Bradovich. The boozers hee-hawed. Bradovich joined them. The judges stopped arguing and raised their fingers like the three monkeys, one to his eyes, one to his ears, one to her mouth.

"I'll not kneel," Bradovich shouted. "I am what I am. Have done what I've done. Will do as I will. Notwithstanding. No fifth amendment for me. And there is no wherefore, therefore, herinafter, *et cetera, et sequens* about it. You can have your *quid pro quos,* your *sine qua nons,* your *res ad judicatas* and who gives a shit?

"What is above is not for us.

"Neither is what is below.

"Nor for us is what will be or what was.

"We did what we did.

"We do what we do.

"We will do what comes naturally."

In the nooks and crannies, the unsheltered, the boozers, the users, awoke, rubbed their phlegmy eyes and brayed. Among the serried rows of stonefaces, there were angry shouts, pointing of fingers, raising of fists. The great clock of the terminal struck 5:30 and passengers began to arrive carrying their valises and backpacks, trains whistled, the loud speaker system proclaimed arrivals and departures, commuter coaches discharged hurrying men and women, a few children, there were comings and goings, bedlam ensued. The judges stood and began packing their briefcases, stonefaces left to begin their daily investigations and surveillances throughout the city, the unsheltered, the boozers, and users stood to stretch, the few who had died during the night curled in rigor mortis.

Bradovich smiled smugly to himself, squared his shoulders, pushed his way through the echoing central hall, ducked under and through the musty corridors, jumped into a waiting cab and rushed home to his apartment where Beatrice Holden awaited him, two stonefaces tailing him closely.

Beatrice greeted him with cheers. "Hurry up, tell me!"

"Later. I want to take a shower. I want to eat a big breakfast. I want to get laid. I want to go to work."

"Okay, Wizard old boy, you go shower, I'll prepare breakfast. Where are you going to eat it? Under the arch?"

Bradovich laughed. She was irrepressible. She was as bouncy as a brand new rubber ball. She was alive.

They ended up naked, entwined, and exhausted on the living room floor, not wanting to chance the early rising Mrs. Kastner at the bedroom window.

"That'll hold me at least a month," he gasped.

"Whatever," she whispered happily.

As they rose to have their second cup of coffee, the telephone rang.

It was Golo the Gimp.

He was hysterical, crying, sobbing, muttering incoherent words about having sold his soul to the devil, and

193

blaming Bradovich, too. Wizard let him go on, every once in a while throwing in a quieting word. After what seemed an hour, Golo stopped babbling, and just wept, intermittently howling like a dog who has lost its master. Bradovich was afraid to ask, but finally gathered up enough courage, terrified at what the answer would be. "Where's Floriana?"

At which Golo began to babble again about his soul, blaming Bradovich, his own friend, cursing Raul, Floriana's brother who had shown up at the arrival gate at the airport, it was them, those bloody bastards, it, The Authority.

Yes, The Authority. It had sent its operatives along on the plane. They were all over, what made the difference? Upon landing, Floriana had coldly turned her back on her brother, and they had gone to their hotel on the beach. Floriana in the bathroom to shower, he'd gone downstairs to buy some Jamaican cigars. When he returned to their suite, he found his lovely Floriana on the floor of the bathroom, her pistol in her hand, her throat slit, her blood wasted.

*C*hallenging The Authority—
it could not be too happy with him after what had happened
in the Grand Old Terminal—Bradovich flew to Guadeloupe,
gathered up the dead Floriana and his hysterical friend and
carried them back to the city. Never once did the Gimp look
him square in the eye.

With great efficiency, Beatrice made all the arrange-
ments. Yesterday a marriage, today a funeral.

Golo had to be sedated, and even then he continued to
babble on about having sold his soul and refused to look
Bradovich in the eye. Wizard no longer listened to him,
merely spoke softly to him, giving him instructions as to his
daily needs, like a father to his child. Bradovich understood
full well what Golo was driving at.

The mortician dressed Floriana in the carmine satin
pants suit which had been her great joy when alive. The slit
throat he neatly sewed together, pancaked and powdered it
so it was imperceptible. Her raven hair he coiffed in the

Spanish style, piled high, and her lips he painted in red matching her suit. She looked her gorgeous voluptuous self ready for bronzing in Miro's metal shop. Her wondrous arch soon to celebrate a victory for ants and worms. Around her luscious thigh Bradovich himself belted her old sheath and dagger. He believed if Floriana had had her dagger instead of the pistol Golo had given her, what happened would not have happened. That dagger had protected her well in life. Now it would protect her against the demons of hell whose existence she had never doubted so help her God. He said nothing of this to Golo.

Beatrice believed one had to mourn honestly, so she neglected to give Golo his sedation. Bradovich then had to use his immense strength to restrain the Gimp from jumping out the window of Martin and Laura's former bedroom where Bradovich had practically imprisoned him. Mrs. Kastner, a witness to Golo's attempt at defenestration, leaned out her window and urged the Gimp on. "Go ahead, jump! Let him go, you big ape. What the fuck you holding him back for? Let him jump! Go ahead, you little jerk, jump!"

Golo almost broke his hold but Bradovich hooked his elbow around Golo's windpipe and squeezed. By natural reflex Golo brought his hands up to pry the elbow loose, and Bradovich then was able to drag him off the window sill, Mrs. Kastner screaming all the time, "Let him jump, you dirty son-of-a-bitch, let him go all the way!"

All the way. Floriana, gorgeous, ridiculous Floriana who'd thought she could beat life, had gone all the way. She'd done what she'd done, fought her way clear, and then The Authority had let her know who was boss. There was to be no returning for lovely Flor.

At the pre-burial rites in the funeral establishment, a block from the park where Floriana had jogged daily, the large chapel was crowded with friends, clients, and the apartment building staff led by slimless Slim. A team of young gymnasts did an acrobatic turn, unaccompanied by music, terminating with a pyramid, a tomb for their Floriana. Still without music, a magician levitated teary-eyed Consuela, then made her disappear into thin air. A gay poet read a mass for the dead. Finally, Madeleine Dearing sang *a capella* Schubert's melancholy *Nacht.*

> . . . .
>
> Soon it will be done,
> Soon I will sleep the long long sleep
> That from all grief frees me. . . .

At the burial site, it took Bradovich and three others to restrain Golo from jumping into Floriana's grave.

Beatrice had had a caterer prepare refreshments for Golo and Floriana's many friends in Bradovich's apartment. The Gimp, no longer sedated, now mourned honestly: he wept, a concerned Consuela constantly at his side. He also drank a lot of bourbon. After the mourning guests left, Golo with Consuela at his elbow, prepared to leave, he would, he said, sleep one last night in Floriana's old apartment (now Consuela's), in the huge walk-in closet fitted out like a stage dressing room. Tomorrow morning, he and Consuela, the new assistant for his magic act, would fly to Tokyo. The show must go on.

He asked Bradovich to sit down, then standing before his friend he at last looked him straight in the eye. His lips were white, and he trembled head to toe. "It was either you or her. It should have been you. She was young, and you're an old bastard. You, yes, it should have been you."

With that, tearful, his small frame quivering, Golo did a shaky about face and hopped gimpily from the room, Consuela behind him.

The apartment door closed, Golo and his new *amie*

gone, Bradovich sat slumped in his chair and looked at Beatrice for comfort. She smiled gently at him. "He's just broken up. He'll soon find succor on Consuela's young tits. It wasn't your fault. As you always say, it was arbitrary. Forget it."

"I don't think he'll ever forgive me. Well, maybe some day, when The Authority goes to work on him, too, and he finds out he's not the exception that proves the rule. We are what we are, we do as we do, it's time to go to work, and I'm beholden to you." He smiled wistfully at her.

Irrepressible Beatrice laughed.

*B*eatrice and Bradovich were married at City Hall.

They didn't take a honeymoon, they felt they didn't need one. Martin and Laura's families and Gregory and Peter, Beatrice's two sons, would come in a month for a big family get-together.

The next day, Beatrice went to her apartment to edit a manuscript about the big bang, quasars, black holes, what have you, and Bradovich walked all the way down to his studio. Unbolted it. Opened up windows and skylights. Aired it out. The wood head in the noose still hung there with the sign "The Authority" tacked on. He left it there. His studio was clean because he always left it clean. He was a fastidious man.

He chose a flawless block of Ferrara marble and working the chain hoist deftly lifted it and placed it in the center of the studio floor under the skylights. There was a young sculptor, whose work he had recently seen, who fascinated him. Made huge forms out of plaster and stone, flexed, torsolike, the essence, you could see life fighting to free itself from within. New, but not new, with history. Back to Rodin, Degas, Michelangelo. Headless. He, Bradovich, would make the heads to go with the torsos. Roughhewn heads struggling to free themselves from the rough stone. He couldn't wait to begin, but first he would do Beatrice's head under a red cloud laced with silver, soon it would rain. With deep-set eyes, a full mouth, and a pointed nose, not too long, but not too short either. Pretty, and yet not so pretty. She would smile, so perhaps the point of the nose would be slightly rounded. The surface would be cross-hatched, not polished.

He chose the chisels he would need, made certain they were sharp, and then the hammers.

He studied the block of marble so hard that he saw it transformed into Beatrice's head and he smiled. He raised the pitching tool, a broad-nosed chisel, and a heavy hammer to begin the rough cut. He struck hard. Clean. His hands were strong, did not quiver.

Bradovich worked all day with the pitching tool, neglecting lunch. He had it all in his fist and he refused to let go of it. Late in the afternoon he phoned Beatrice. "How you doing with your work?"

"Fine. Do you know the universe never stops expanding? Never! It's infinite up there. Endless. It's awesome. Makes me scared. What about you, how did it go?"

"Great. So good I don't want to stop. What about coming down here? We'll go to the riverfront diner for a bite, then return here so I can work some more."

"I'll come, but I'll bring the food since it's already made."

"Good. Then you can sit for me."

"Why? Haven't we done it sitting yet?"

"Bring a book to read."

"See you soon, Bradovich."

She arrived an hour later by cab, with a hot veal stew, a green salad with oil and vinegar dressing and French bread. Bradovich made coffee on his hot plate.

After they ate and cleaned up, he showed her about his studio, the different kinds of marble, the lengths of wood. She loved a small block of dyed Swiss pear. "Okay, someday I'll carve you a miniature laughing head. It will be just another piece in Bradovich's Beatrice period."

"Are you going to keep that there?" she said, pointing to the wood head in the noose.

"Yes," he said. "Fear feeds fear. Courage breeds courage. I'm not afraid of them."

She laughed. "You're okay, Mister Bradovich." Then she remembered that was Valerian's. "I'm sorry."

"No, no. You can call me any goddamned thing you want. Val is Val. You're you. I have no problem with it."

He brought her an armchair. "Sit here. You don't have to pose, just relax, read your book if you want. What is it, by the way?"

"Kafka," she said.

"I ought to make a head of him."

"Perhaps you should."

They didn't speak much as Bradovich worked. A word here or there about what she'd just read, about Golo.

"He'll probably marry Consuela," Beatrice said.

"Most likely. Why, I don't know, since he's always chasing after pussy. Still, he's dying to father a child, perhaps that's why."

Bradovich liked having her in his studio. It was a good feeling, it made him whole again. He worked quickly and was already through with the pitching tool, the form of the head roughly defined. Now he was working with the boaster and iron mallet, removing a second layer of stone. "It's like peeling an artichoke," he told her, "leaf by leaf until you get to the heart."

But soon he forgot her presence, his concentration so profound there was nothing in the world but the hard piece of marble that was yielding to his persistent hands. When, finally tired, he looked up, he found her staring at him through sleepy eyes, a gentle bemused smile on her lips.

"Let's sleep here tonight," he said. "In the morning, we'll go to the diner for breakfast."

"Sounds great."

"The bathroom's there," he pointed. "It's been renovated, and there's plenty of hot water."

As they rested comfortably on the foam slab on the floor, nestled sleepily in each other's arms, he whispered, "We'll have a good life together."

"Yes," she said huskily. "You'll do your work, and I'll do mine. I'll analyze books to death for you, and in the galleries you'll point out the difference between the fake that you're always yelling about and the true of which there is never enough. We'll pester our children until they're glad to be rid of us. We'll do okay."

"We'll do what is rightly done," Bradovich said, as they settled down to sleep.

Outside the studio, in the quiet autumn night, lolled two stony-faced operatives dressed in black *cap-à-pie*, except for starched white shirts, button-down. The Authority, arbitrary though it is, has great patience.